Letts study aids

Revise Physics

A complete revision course for O level and CSE

Michael Shepherd BSc, PhD, MInstP

Senior Science Master, Malvern College, Worcestershire

Charles Letts Books Ltd
London, Edinburgh & New York

First published 1979
by Charles Letts & Co Limited
Diary House, Borough Road, London SE1 1DW
Reprinted 1980
Revised 1981
Reprinted 1982 (twice)
Design: Ben Sands
Illustrations: Rod Fraser

ISBN 0 85097 397 X

Printed by Charles Letts (Scotland) Limited

Preface

This book is intended to be a complete guide for the student revising for the GCE 'O' level, SCE or CSE examination in Physics. It is primarily intended for use during the eight months or so prior to the examination, but may also prove useful at other times during the course.

Revise Physics differs in many respects from a conventional text book. Firstly it is based on a thorough analysis of the Physics syllabuses of the GCE 'O' level, SCE and CSE Examination Boards, and, unlike a conventional text book, informs the student of the scope of a particular syllabus for which he or she is preparing. (The table of analysis clearly displays how the requirements of each Board relate to the core revision material.) Secondly, the core material consists of a detailed and concisely expressed account of Physics to this level in a form most suitable for revision. A further feature of this book is the inclusion of a large number of diagrams, for so much Physics which is difficult to explain in words, can be shown clearly in diagrams.

Following the core material there is a section on examination technique, which contains advice for the student on how to tackle the various types of examination papers and the questions they contain. This is followed by a self-test section containing many multiple choice questions. These are designed to enable the student to judge his or her progress; answers are provided. After the self-test section there is a large selection of questions taken from past papers of the different Examination Boards. Multiple choice, short answer and longer questions are included. Each longer question is followed by an outline of the physical principles involved in answering it.

I am extremely grateful to Gwen Mossop and Bob Repper for their assistance in the preparation of this book. Their work and comments have been much appreciated. I would also like to thank Mrs Pat Turton for typing the manuscript with such patience and accuracy, and the staff of Charles Letts & Co Limited for their valuable advice and unfailing courtesy. In the revision of this book I am indebted to Jim Wilson for guidance on the Scottish material.

I also wish to thank the many Examination Boards who have granted permission for the reproduction of examination questions taken from their papers. The names and addresses of all the Examination Boards and the abbreviations used for them in this book are given after the table of analysis.

Michael Shepherd 1981

Contents

Section I
Introduction and guide to using this book

Knowing how to prepare for an examination can be difficult. *Revise Physics* will make it easier by providing the GCE 'O' level, SCE and CSE Physics student with a complete revision course. This book not only contains all the necessary topics, but has a self-test section and an extensive selection of actual past examination questions with answers. Each longer question is keyed back to the relevant unit in the book and is followed by hints on how to answer it. To get the best from the study scheme in this book you are advised to follow the procedure outlined below.

(a) **Using the table of analysis of examination syllabuses.** Turn to pages 2 and 3 for the table of analysis. The Examination Boards (and their alternative syllabuses where relevant) are listed across the top of each page. The contents of this book are listed by unit and sub-unit headings down the left hand side of the page.

Find the column containing the information about the syllabus you are studying. (Remember, many Examination Boards have alternative Physics syllabuses, so if you are in doubt about the requirements of your particular Board, please check with your teacher – the table is intended only as a useful guide.) In the column you will find the following information:

  (*i*) the number of examination papers set with the time allowed in brackets.

 (*ii*) whether there is a practical examination. (The figure in brackets indicates the percentage of the total mark allocated to the practical examination.)

(*iii*) whether there is any teacher assessment. (This is a mark awarded by your teacher for the work you have done at school.)

(*iv*) whether multiple choice and free response (short answer, structured or essay) questions are set. The figures in brackets give the allocation of marks to these different types of question.

 (*v*) The Physics content of the GCE, SCE and CSE courses – the most important information. This information has been divided into 28 units, within which are a varying number of sub-units. Not all of these sub-units need to be studied for any one syllabus. The key to the symbols in the columns is:

    ● unit required for a syllabus

    ○ unit optional for a syllabus

    ●/○ unit required as part of a syllabus but also required in greater depth as an option.

A blank space indicates that a unit is not required.

(b) **Section II: the core material.** You are advised to work through as many of the required units as possible as well as some of the optional units. Your teacher will advise you which options you should study. You may still find it useful to consult your Examining Board's syllabus as it may give valuable information in the form of additional notes. You can write to the Boards to buy syllabuses and past papers. A list of their addresses is given on page 8.

(c) **Section III: examination technique.** This part of the book gives you advice on tackling the different kinds of examination questions and how to do your best under examination conditions.

(d) **Section IV: testing yourself and revising.** When you feel you have mastered as many units as you can, you should test what you have learned. A self-test section, consisting of multiple choice questions, has been included to enable you to judge your own progress. If you fail to get the majority of these right, go back to the appropriate unit in the core material and revise more thoroughly.

(e) **Section V: practice in answering examination questions.** A large selection of questions from past examination papers has been included to give you maximum practice in answering all types of questions. For convenience these questions have been grouped by type and each type subdivided according to whether the questions are more suitable for the GCE 'O' level, SCE or CSE examination. However GCE candidates may benefit from trying them all. Be careful not to answer questions on sub-units of the core material which are not in your syllabus.

Table of analysis of examination syllabuses

	AEB	Cambridge	JMB	London	O and C	Oxford 5857	Oxford 5858	SUJB	WJEC	Nuffield (O and C)	SCE	ALSEB
Level	O	O	O	O	O	O	O	O	O	O	O	CSE
Syllabus						5857	5858					
Number of papers + (total time)	$3(3\frac{3}{4})$	$2(4\frac{1}{2})$	$1(2\frac{1}{2})$	$2(3\frac{1}{4})$	$2(4)$	$2(3\frac{1}{2})$	$3(4\frac{1}{2})$	$1(2\frac{1}{2})$	$1(2\frac{1}{2})$	$2(3\frac{1}{2})$	$2(2\frac{1}{2})$	$2(2\frac{1}{4})$(a
Practical examination												●25%
Teacher assessment % marks						●10%						●15%
Multiple choice	●30%		●30%	●40%			●35%			●50%	●56%	●35%(
Free response	●70%	●100%	●70%	●60%	●100%	●100%	●55%	●100%	●100%	●50%	●44%	●25%
1 Measurements (all)	●	●	●	●	●	●	●	●	●	●	●	●
2.1 Tickertape vibrator (1)												
2.2 Motion	●	●	●	●	●	●	●	●	●	●	●	●
2.3 Speed	●	●	●	●	●	●	●	●	●	●	●	●
2.4 Velocity	●	●	●	●	●	●	●	●	●	●	●	●
2.5 Acceleration	●	●	●	●	●	●	●	●	●	●	●	●
2.6 Uniformly accelerated motion			●	●		●	●	●	●	●(2)	●	●(3)
2.7 Distance travelled				●		●	●	●	●	●(2)		●(3)
2.8 Experiments to show uniform acceleration	●	●	●	●	●	●	●	●	●	●	●	●
2.9 Projectiles				●						●	●	
3.1 Newton's laws of motion	●	●	●	●	●	●	●	●	●	●		●(4)
3.2 The acceleration produced by a force	●	●	●	●	●	●	●	●	●	●		
3.3 Weight	●	●	●	●	●	●	●	●	●	●	●	●
3.4 Motion in a circle		●(5)		●(5)		●(6)	●	●(6)	●(6)	●		
3.5 Momentum	●		●	●	●	●	●	●(6)	●	●(2)	●	●(3)
3.6 Action and reaction	●		●	●	●	●	●	●(6)	●			●(3)
3.7 Work	●	●	●	●	●	●	●	●	●	●	●	●
3.8 Energy	●	●	●	●	●	●	●	●	●	●	●	●
3.9 Potential energy	●	●(6)	●	●	●	●	●	●(6)	●	●(2)	●	●(3)
3.10 Kinetic energy	●	●(6)	●	●	●	●	●	●(6)	●	●(2)	●	●(3)
3.11 Power	●	●	●	●	●	●	●	●	●	●	●	●
4.1 Scalar and vector quantities	●	●		●	●	●	●	●	●	●	●	●
4.2 The parallelogram of forces	●	●		●	●	●	●	●	●(7)	●	●	●
4.3 Resolution of forces				●	●	●	●	●	●(7)		●	
4.4 The triangle of forces	●											
5.1 Moments	●	●		●	●	●	●	●	●	●	●	●(3)
5.2 Parallel forces	●	●		●	●	●	●	●	●	●		●(3)
5.3 Centre of mass	●	●		●	●	●	●	●	●		●	●
5.4 Location of the centre of mass	●			●	●	●	●	●	●			●
5.5 Stability	●			●	●	●	●	●	●		●	●
5.6 Machines	●			●	●	●	●	●	●		●	●
5.7 Work done by a machine	●			●	●	●	●	●(8)	●		●	●
5.8 The lever	●			●	●	●	●	●	●(10)	●	●	●
5.9 Pulleys	●		●	●	●	●	●	●	●		●	●
6 Density	●	●	●	●	●	●	●	●	●	●		●
6.1 Density measurements	●	●	●	●	●	●	●	●	●	●		●
6.2 Relative density			●	●				●		●		
7 Pressure	●	●	●	●	●	●	●	●	●	●		●
7.1 Pressure in a fluid	●	●	●	●	●	●	●	●	●	●		●
7.2 Atmospheric pressure	●	●	●	●	●	●	●	●	●	●	●	●
7.3 The manometer	●	●(6)	●	●	●	●	●	●		●		●

Key

(a) An additional ten minutes are allowed for reading Paper 2.
(b) Mainly, but not exclusively, objective type (multiple choice) questions.
(c) A Nuffield based syllabus (B) is alternative to syllabus N or syllabus S. Paper 1 ($1\frac{1}{4}$ hours) consists of 50 multiple choice questions carrying 35% of the marks. Paper 2 ($2\frac{1}{4}$ hours) comprises seven structured questions carrying 35% of the marks. Teacher assessment of practical work carries the remaining 30% of the marks.
(d) An additonal ten minutes are allowed for reading each paper.
(e) Teacher assessment of practical work is an alternative to the practical examination.
(f) A Nuffield based course.
(g) An additional fifteen minutes are allowed for reading Paper 2.
(h) No information is available.
(j) Includes a $\frac{3}{4}$ hour paper (15%) which may be examined as an alternative by a candidate's teacher.
(k) Includes 15% for either an optional topic (teacher assessed) or for Paper 2 testing four optional topics.

(l) 40% of the marks are allocated for practical questions done in the last two terms and assessed by the teacher.
(m) A Nuffield based course is available as an alternative. The syllabus is that shown in the Nuffield column, but excluding part of the quantitative kinetic theory, part of modern Physics, and all of the mechanical equivalent of heat and astronomy.
(n) A Nuffield based course. Circular motion and Planetary Astronomy are excluded.
(p) An alternative is a test of practical knowledge by means of questions requiring written answers.

(1) Use only required.
(2) The equations of motion are printed on the examination paper.
(3) Further Physics alternative only.
(4) A formal statement is not required.
(5) Knowledge of the equation $F = mv^2/r$ is not required.
(6) Qualitative only.
(7) Vectors at right angles only.

EAEB		EMREB		LREB		NREB	NWREB		SREB	SEREB	SWEB	WJEC	WMEB	WY & LREB	YREB	
CSE	CSE	CSE	CSE	CSE	CSE	CSE	CSE	CSE	CSE	CSE	CSE	CSE	CSE	CSE	CSE	
North(c)	South(c)	1	2	A	B(f)		A	B	(n)	(m)						
2(3¼)	2(3¼)	2(3)(d)	2(3)(d)	2(2¼)	2(2¼)	↑	2(2¼)(j)	1(2½)	1(2⅙)	2(2½)(a)	2(3)	1(2½)	2(4)	2(2½)(a)	2(3)	
		●30%(e)	●30%(e)	●20%	●20%		●20%	●25%(e)	●20%(l)		●25%					
		(e)	(e)			(h)	●15%			●60%			●20%	●20%	●(p)	
●50%	●40%			●25%	●25%		●30%	●35%	●40%	●15%				●40%	●40%	
●50%	●60%	●70%	●70%	●55%	●55%	↓	●35%(k)	●40%	●40%	●25%	●75%	●100%	●80%	●40%	●60%	
●	●	●	●	●	●	●	●	●	●	●	●	●	●	●	●	1
																2.1
●		●	●	●	●	●	●	●	●	●	●	●	●	●	●	2.2
●	●	●	●	●	●	●	●	●	●	●	●	●	●	●	●	2.3
●	●	●	●	●	●	●	●	●	●	●	●	●	●	●	●	2.4
●	●	●	●	●	●	●	●	●	●	●	●	●	●	●	●	2.5
●					●								●	●	●	2.6
●					●										●	2.7
●	●	●	●	●	●	●	●	●	●	●		●	●	●	●	2.8
				●	●											2.9
●(4)	●(4)	●(4)	●(4)	●(4)	●	●(4)	●(4)	●(4)	●(4)	●(4)	●	●	●	●	●	3.1
●	●	●	●	●	●	●	●	●	●	●	●	●	●	●	●	3.2
●	●	●	●	●	●	●	●	●	●	●	●	●	●	●	●	3.3
					●								●(6)			3.4
				●(2)					●						●(6)	3.5
		○			●				●							3.6
●	●	●	●	●	●	●	●		●	●	●	●	●	●	●	3.7
●	●	●	●	●	●	●	●		●	●	●	●	●	●	●	3.8
●(6)	●	●	●	●	●(2)	●	●	●	●(6)	●	●(6)		●(6)	●(6)	●(6)	3.9
●(6)	●	●	●	●	●(2)	●	●	●	●(6)	●	●(6)		●(6)	●(6)	●(6)	3.10
●(6)	●	●	●	●	●	●	●	●	●	●	●		●	●	●	3.11
●	●	○			●	●							●	○		4.1
●(7)	●	○				●								○		4.2
		○							●					○		4.3
		○												○		4.4
●	●	●	●	●			●	●	●	●	●		●	●	●	5.1
●	●	●	●	●			●	●	●	●			●	●	●	5.2
●	●	●		●		●	●	●	●	●	●			●	●	5.3
●	●	●		●		●	●	●	●	●	●			●	●	5.4
	●	●		●			●		●	●					●	5.5
●	●	○		○		●	●	●	●	●	●	●	●	●/○	●	5.6
●	●	○		○		●	●	●	●	●(9)	●	●	●	●	●	5.7
●	●	○	●	●/○		●	●	●	●	●	●	●	●	●/○	●	5.8
●	●	○	●	○		●	●	●	●	●	●	●	●	●/○	●	5.9
●	●	●	●	●		●	●	●	●	●	●	●	●	●	●	6
●		●	●	●		●	●	●	●	●	●		●	●	●	6.1
				●			●	●		●	●		●	●	●	6.2
●	●	●	●	●	●	●	●	●	●	●	●	●	●	●	●	7
●	●	●	●	●	●	●	●	●	●	●	●	●	●	●	●	7.1
●(11)	●	●	●	●	●	●	●	●	●	●	●	●	●	●	●	7.2
●(11)	●	●	●	●	●	●							●		●	7.3

(8) The terms 'mechanical advantage' and 'velocity ratio' will not be used.

(9) Efficiency in terms of work only.

(10) Wheel and axle only.

(11) Knowledge of the variation of pressure with depth is required.

(12) Knowledge expected but not tested explicitly.

(13) Detailed description not required.

(14) Idea of molecular activity.

(15) Questions that require knowledge of a particular thermometer will not be set.

(16) The theoretical and experimental basis of the ideal gas equation will not be examined.

(17) Not the experimental determination.

(18) Plane boundaries only.

(19) Plane waves only.

(20) Uses only.

(21) Knowledge of eye defects and their correction is not required.

(22) Knowledge of one simple experiment to establish the wave theory of light is required.

(23) Not charging by induction.

(24) Equal resistors only.

(25) Not the rms value.

(26) Semiconducting.

(27) Knowledge of one detector is required.

Table of analysis of examination syllabuses – *continued*

		AEB	Cambridge	JMB	London	O and C	Oxford	SUJB	WJEC	Nuffield	SCE	ALSEB
7.4	Simple barometer	●	●	● (12)	●	●	●	●	●	●		●
7.5	Aneroid barometer					●	●					●
8.1	Archimedes' principle	●	● (6)		●	●	●	●	●			●
8.2	Theory					●	●					●
8.3	Balloons	●	●		●	●	●	●		●		
8.4	Ships	●	●		●	●	●	●		●		
8.5	Hydrometers	●	●		●	● (13)	●	●	●			●
9.1	Elasticity	●		●	●		●			●		
10	Surface tension					●	●	●	●			
11.1	Atoms and molecules	●	●		●	●	●	●	●			● (3)
11.2	The states of matter	●	●	●	●	●	●	●	●	●	●	●
11.3	Brownian motion	●	●	●	●	●	●	●	●	●	●	
11.4	Diffusion	●	●	●	●	●	●		●	●	●	
11.5	Randomness and mean free path	●	●	●	●	●	●	●	●		●	● (3)
12.1	Solids	●	●	● (12)	●	●	●	●	●	●		●
12.2	Expansivity	●		● (12)	●	●	●	●				● (3)
12.3	Liquids	●	●	● (12)	●	●	●	●	●	●		●
12.4	Thermometers	●	●	● (12)	●	●	●	● (15)	●	● (15)	●	●
13.1	Boyle's law	●	●	●	●	●	●	● (16)	●	●	●	● (3)
13.2	Charles' law	●	●	●	●	●	●	● (16)	●	●	●	● (3)
13.3	Pressure law	●	●	●	●	●	●	● (16)	●	●	●	● (3)
13.4	The universal gas law	●	●	●	●	●	●	●		●		● (3)
13.5	Models of a gas	●	●	●	●	●			●	●		● (3)
14	Specific heat capacity	●	●	●	●	●		●	●	●		●
14.1	The specific heat capacity of a solid	●	●		●	●		●	●	●		●
14.2	The specific heat capacity of a liquid	●	●		●	●		●	●	●		●
15.1	Vaporisation	●	●	●	●	●		●	●	●		●
15.2	Fusion	●	●	●	●	●		●	●	●		●
15.3	The specific latent heat of fusion of ice	●	●	● (17)	●	●	●		●	●		●
15.4	The specific latent heat of vaporisation of water	●	●	● (17)	●	●	●	●	●	●		●
15.5	The effect of impurities	●	●		●	●						●
15.6	The effect of pressure on the melting temperature of ice	●	●		●	●		●				●
15.7	The effect of pressure on the boiling temperature of water	●	●		●	●		●				●
15.8	Evaporation	●	●	●	●	●	●	●	●	●	●	○
15.9	Vapour pressure	●			●	●	●	●	●			
16.1	Conduction	●	●	●	●	●	●	●	●	●		●/○
16.2	Comparison of thermal conductivities	●	●	●	●	●	●	●		●		●/○
16.3	Radiation	●	●	●	●	●	●	●	●	●		●
16.4	Convection	●	●	●	●	●	●		●	●		●/○
17.1	Progressive waves	●	●	●	●	●	●	●	●	●		●/○
17.2	The ripple tank	●	●	●	●	●	●	●	●	●	●	●
17.3	Reflection	●	●	●	●	●	●	●	●	●	●	●
17.4	Refraction	●	●	●	●	●	●	●	●	●	●	●
17.5	Interference		●	●	●	●	●	●	●	●	●	●
17.6	Diffraction		●	●	●	●	●	●	●	●	●	
17.7	3 cm waves				●			●		●	●	
18.1	Rectilinear propagation	●	●	●	●	●	●	●	●	●		
18.2	Reflection at a plane surface	●	●	●	●	●	●	●	●	●	●	●
18.3	Reflection at a curved surface	●		●	●	●	○	●	●			●
18.4	Construction of ray diagrams	●		●	●	●		●	●			●
19.1	Refraction at a plane boundary	●	●	●	●	●	●	●	●	●	●	●
19.2	Internal reflection and critical angle	●	●		●	●	●	●	●	●	●	●
19.3	Refraction at a spherical boundary		●	●	●	●	●	●	●	●		●
19.4	The focal length of a converging lens	●	●	●	●	●	●	●	●	●		●
19.5	Construction of ray diagrams	●	●	●	●	●	●	●	●	●		●
19.6	Simple microscope	●	●	●	●			●	●	●		
19.7	The projector	●	●			●	○					
19.8	The camera	●	●	●	●	●	○	●	●	●		● (3)
19.9	The eye		●	●	●	● (21)	○	●	●	●		● (3)
19.10	The compound microscope						○	●	●	●		
19.11	The telescope			●			○	●	●	●		○
20.1	Diffraction		● (22)	●	●	●	●	●		●	●	
20.2	Interference		● (22)	●	●	●	●	●		●	●	
20.3	The diffraction grating		● (22)			●	●	●		●	●	
20.4	The electromagnetic spectrum	●	●	●	●	●	●	●	●	●	●	●

EAEB		EMREB		LREB		NREB	NWREB		SREB	SEREB	SWEB	WJEC	WMEB	WY & LREB	YREB	
● (11)	●	●/○	●	●	●	●	●	●	●	●	●	●	●	●/○	●	7.4
	●	●/○		●		●	●	●		●	●			●/○	●	7.5
●	●	●		●		●	●	●		●	●	●		●		8.1
																8.2
●	●	●					●	●				●				8.3
●	●	●		●		●	●	●	●	●		●		●		8.4
●	●	●		●		●	●	●	●	●		●		●	●	8.5
●	●	●	●	●	●	●			●	●	●	●	●	●	●	9.1
	●	●	●	●			●	●				●	●			10
●	●	●	●	●	●	●	●	●	●	●	●	●	●	●	●	11.1
●	●	●	●	●	●	●	●	●	●	●	●	●	●	●	●	11.2
● (14)	● (14)	●	●	●	●		●	●	●	●	●	●	●	●	●	11.3
● (14)	● (14)	●	●	●	●				●	●	●	●	●	●	●	11.4
					●											11.5
● (12)	●	●	●	●		●	●	●	●	●	●	●	●	●	●	12.1
	●			●		●		●					●			12.2
● (6)	●	●	●	●		●	●	●	●	●	●	●	●	●	●	12.3
●	●	●	●	●/○		●	●	●	●	●	●	●	●	●	●	12.4
●	●			●	●	●	● (6)		●		● (6)	● (6)	● (6)	●	●	13.1
● (6)	●	● (6)	●	●		●	● (6)		●	● (6)	● (6)	● (6)	● (6)	● (6)	●	13.2
● (6)	●	● (6)		●			● (6)		●	● (6)	● (6)	● (6)	● (6)	● (6)	●	13.3
							● (6)			● (6)	● (6)	● (6)				13.4
●	●	●	●	●	●		●	●	●	●	●	●	●	●	●	13.5
●	●	●		○	●	●	●	●	●	●	●	●	●	●	●	14
	●	●		○	●	●	●	●	●	●	●	●	● (6)	●	●	14.1
	●	●		○	●	●	●	●	●	●	●	●	● (6)	●	●	14.2
●	●	●	●	●	●	●	●	●	●	●	●	●	●	●	●	15.1
●	●	●	●	●	●	●	●	●	●	●	●	●	●	●	●	15.2
● (6)	● (6)	● (17)		○		●	●	●	●	●	●	●		●	●	15.3
● (6)	● (6)	● (17)		○		●	●	●	●	●	●	●		●	●	15.4
						●	●	●			●			●	●	15.5
						●	●	●			●			●	●	15.6
●				●		●	●	●	●	●	●		●	●	●	15.7
●		●/○	●	●/○	●	●	●	●	●		●	●	●	●		15.8
																15.9
●	● (6)	●	●	●	●	●	●	●	●	●	●	●	●	●	●	16.1
●	● (6)	●	●	●	●	●	●	●	●	●	●	●	●	●	●	16.2
●	● (6)	●/○	●	●	●	●	●	●	●	●	●	●	●	●	●	16.3
●	● (6)	●	●	●	●	●	●	●	●	●	●	●	●	●		16.4
●	●	●	●	●	●	●	●	●	●	●	●	●	●	●		17.1
●	●	●	●	●	●					●	●	●	●	●		17.2
● (18)	●	● (18)	●	●	●					●	●	●		● (20)		17.3
● (18)	●	● (18)	●	●	●					●	●	●		● (20)		17.4
●			●	●	●					●						17.5
●			●	●	●					●		●		● (19)		17.6
										●						17.7
●	●	●/○	●	●	●	●	●	●	●	●	●	●	●	●	●	18.1
●	●	●	●	●	●	●	●	●	●	●	●	●	●	●	●	18.2
● (20)	●			●/○		●	●	●	●	●	●	●		●	●	18.3
				●			●	●			●					18.4
●	●	●	●	●	●	●	●	●	●	●	●	●	●	●	●	19.1
●	●				●	●	●	●	●	●	●	●			●	19.2
●	●		●	●/○	●	●	●	●	●	●	●	●	●	●	●	19.3
●	●		●	●	●	●	●	●	●	●	●	●	●	●	●	19.4
●	●		●	●	●	●	●	●	●	●	●	●	●	●	●	19.5
●	●	●		○	●	●	○			●	●				●	19.6
	●	○		●	●	●	○			●	●			○	●	19.7
●	●	○	●	○	●	●	●/○		●	●	●	●	●	●/○	●	19.8
●	●	○	●	○	●	●	●	●	●	●	●	●	●	●	●	19.9
		○	●	○	●		○									19.10
	●	○	●	○	●	●	○	●	●	●			○		●	19.11
				●	●											20.1
				●	●											20.2
					●											20.3
●	●	●	●	○	●	●	●	●	●	●	●	●	●	●		20.4

Table of analysis of examination syllabuses – *continued*

	AEB	Cambridge	JMB	London	O and C	Oxford		SUJB	WJEC	Nuffield	SCE	ALSEB
21 Sound waves	●	●	●	●	●	●	●	●	●		●	●
21.1 Echoes	●	●		●	●	●	●	●	●		●	●
21.2 Pitch	●	●	●	●	●	●	○	●	●		●	●
21.3 Intensity and loudness	●	●		●	●		○	●	●		●	
21.4 Quality	●	●		●	●		○	●	●		●	
21.5 Interference of sound				●		●						
21.6 Stationary waves	●		●	●	●		○	●	●			
21.7 The sonometer	●		●	●	●		○	●				●
21.8 Forced vibrations	●		●	●	●		○	●	●			
21.9 Resonance	●		●	●	●		○	●	●			
22 Magnets	●	●	● (12)	●	●	●	●	●	●		●	●
22.1 Making magnets	●	●	● (12)	●	●	●	●	●	●		●	●
22.2 Demagnetisation	●	●	● (12)	●	●	●	●	●	●			●
22.3 Magnetic fields	●	●	● (12)	●	●	●	●	●	●			●
23 Electrostatics	●	●	●	●	●	●	●	●	●		●	●
23.1 The gold-leaf electroscope	●		●	●	●	●	● (23)	●	●		●	
23.2 Distribution of charge over the surface of a conductor			●	●				●	●		●	
23.3 Point action			●	●				●			●	
23.4 The Van de Graaff machine			●								●	
23.5 Lines of force			●	●	●			●	●	●	●	
23.6 Potential			●	●	●		●	●	●		●	
23.7 Capacitance			●	●		●	●	●	●		●	
23.8 Capacitors			●	●		●	●	●				○
24 Electric current	●	●	●	●	●	●	●	●	●	●	●	●
24.1 Potential difference	●	●	●	●	●	●	●	●	●	●	●	●
24.2 Electromotive force	●	●		●	●	●	●	●				
24.3 Resistance	●	●	●	●	●	●	●	●	●	●	●	●/○
24.4 Ammeters and voltmeters	●	●	●	●	●	●	●	●	●	●	●	
24.5 Resistors in series	●	●	●	●	●	●	●	●	●	●	●	● (3)/C
24.6 Resistors in parallel	●	●	●	●	●	●	●	●	●	● (24)	●	● (3)/C
24.7 Energy	●	●	●	●	●	●	●	●	●	●	●	●
24.8 Power	●	●	●	●	●	●	●	●	●	●	●	●
24.9 Cost	●										●	●
24.10 House electrical installation	●	●	●	●				●	●		●	○
24.11 Fuses	●	●	●	●				●	●	●	●	●
24.12 Earthing	●	●	●	●				●	●		●	●
25 Electromagnetism	●	●	●	●	●	●	●	●	●		●	●
25.1 The electromagnet	●	●		●	●	●	●	●	●		●	
25.2 The electric bell			●		●			●	●			●
25.3 Moving-iron instruments			●					●				
25.4 The magnetic field due to a current in two parallel wires	●	● (6)	●	●	●	●	●	●	●		●	● (3)
25.5 The force on charges moving in a magnetic field	●	●	●	●	●	●	●	●	●	●	●	
25.6 The d.c. electric motor	●	●	●	●	●	●	●	●	●		●	
25.7 The moving coil meter	●	●		●	●	●	●	●				● (3)
25.8 Ammeters and voltmeters	●	●		●	●	●	●					● (3)
26.1 The laws of electromagnetic induction	●	●	●	●	●	●	●	●			●	● (3)
26.2 The simple d.c. dynamo	●	●		●	●	●	●	●			●	
26.3 The simple a.c. dynamo	●	●		●	●	●	●	●	●	●	●	● (3)
26.4 Alternating current	●		●	●				●	●		●	
26.5 The transformer	●	●	●	●				●	●	● (25)	●	○
26.6 Power transmission	●	●	●	●		●		●	●	●	●	●
27.1 Thermionic emission	●	●	●	●	●	●		●		●	●	
27.2 The diode	●	●	●	●	●	●	●	●	●	●	●	● (3)
27.3 Cathode rays (electron beams)	●	●	●	●	●	●	●	●	●	●	●	● (3)/C
27.4 The effect of electric and magnetic fields	●	●	●	●	●	●	●	●	●	●	●	● (3)/C
27.5 Energy and velocity of electrons	●										●	● (3)/C
27.6 Cathode ray oscilloscope	●										●	
28.1 Radiation detectors	●	●	●	●	●	●	●	●	●	●	●	● (3)/C
28.2 Atomic structure	●	●	● (27)	●	● (27)	●	●			●		○
28.3 Isotopes	●	●	●	●			●	●			●	● (3)/C
28.4 Radioactivity	●	●	●	●	●	●	●	●			●	● (3)
28.5 Radioactive decay	●	●	●	●	●	●	●	●	●	●	●	○
28.6 Safety		●	●	●	●	●			●	●	●	● (3)/C

EAEB		EMREB		LREB		NREB	NWREB		SREB	SEREB	SWEB	WJEC	WMEB	WY & LREB	YREB	
	●	●	●	●		●	●	●	●	●	●	●	●	●	●	21
	●	●	●	●		●	●	●	●	●	●	●	○		●	21.1
	●	○		○		●	●	●	●	●	●	●	○	●/○	●	21.2
	●	○		○		●	●	●					○	●/○	●/○	21.3
	●	○		○		●	●	●	●					○	○	21.4
		●														21.5
		○		○			●	●				●	○	○	○	21.6
		○		○			●	●				●	○	○	○	21.7
		○		○		●	●	●	●	●			○	○	○	21.8
		○		○		●	●	●	●	●				○	○	21.9
●	●	●	●		●	●	●	●	●	●	●	●	●	●	●	22
●	●	●					●		●		●	●	●		●	22.1
●	●	●					●		●				●		●	22.2
●	●	●			●	●	●	●	●	●	●	●	●	●	●	22.3
●	●	●			●	●	●	●	●	●		●	●	●		23
		●			●	●		●					●			23.1
					●											23.2
					●											23.3
					●			●								23.4
		●			●			●								23.5
		●			●	●	●	●								23.6
				○										○		23.7
				○										○		23.8
●	●	●	●	●	●	●	●	●	●	●	●	●	●	●	●	24
●	●	●	●	●	●	●	●	●	●	●	●	●	●	●	●	24.1
		●	●		●	●	●								●	24.2
●	●	●	●	●	●	●	●	●	●	●	●	●	●	●/○	●	24.3
●	●	●	●	●	●	●	●	●	●	●	●	●	●	●	●	24.4
●	●	●		●	●	●	●	●	●	●	●	●	●		●	24.5
●	●	●		●	● (24)	●	●	●	●	●	●	●	●		●	24.6
●	●	●	●	●	●	●	●	●	●	●	●	●	●	●	●	24.7
●	●	●	●	●	●	●	●	●	●	●	●	●	●	○	●	24.8
○		●			●			●			●	●	●	●	●	24.9
●	●	○				●	○		●	●	●	●	●	○	○	24.10
●/○	●	●	●	●	●	●	●/○	●	●	●	●	●	●	●/○	●	24.11
●/○	●	●	●	●	●	●	●/○	●	●	●	●	●	●	●/○	●	24.12
●	●	○	●	●	●	●	●	●	●	●	●	●	○	●	●	25
●	●		●	●/○	●	●	●	●		●					●	25.1
●	●			○		●	●	●	●		●		○	●		25.2
●				○			●			●	●				●	25.3
																25.4
		●	●	○	●	●	●	●	●	●	●					25.5
●	●	●	●	○	●	●	●	●	●	●	●	●	○	●	●	25.6
●	●	○	●	○	●	●	●	●	●	●	●		○	●	●	25.7
	● (6)	○							● (6)	●	● (6)				● (6)	25.8
●	●	●	●	○		● (6)	● (6)	● (6)	● (6)	● (6)	● (6)	●	○ (6)	● (6)	● (6)	26.1
					●	●	●	●	●	●	●	●	○	●		26.2
●	●	●	●			●	●	●	●	●	●	●	○		●	26.3
●		○	●	○	● (25)					●					●	26.4
●	●	●	●	○	●	●	●	●	●	●	●		●	●	●	26.5
●		○	●	○	●		○		●	●	●		●	○	●/○	26.6
	●	○	●	○	●		○	●	●	●	●			●		27.1
	●	○	●	○	●		○	●	● (26)	●	●			●	●	27.2
	●	○	●	○	●		○	●	●	●	●			●	●	27.3
	●	○	●	○	●		○	●	●	●	●			●	●	27.4
	●	○					○									27.5
	●	○		○		●	●	●	●	●	●		○	●/○	●	27.6
	●	○	●	○	●	●	○	●	●	●			●		●	28.1
●	●	○	●	○	●		●	●	●	●			○		●	28.2
●	●	○	●				○	●		●	●		○		●	28.3
●	●	○	●	○	●	●	○	●	●	●	●	●	●/○	●	●	28.4
●		○	●	○	●		○	●	●	●			○		●	28.5
●	●	○	●	○	●	●		●	●	●			●		●	28.6

EXAMINING BOARDS

GCE Boards

AEB	Associated Examining Board Wellington House, Aldershot, Hampshire GU11 1BQ
Cambridge	University of Cambridge Local Examinations Syndicate Syndicate Buildings, 17 Harvey Road, Cambridge CB1 2EU
JMB	Joint Matriculation Board Manchester M15 6EU
London	University Entrance and School Examinations Council University of London, 66–72 Gower Street, London WC1E 6E
Oxford	Oxford Local Examinations Delegacy of Local Examinations, Ewert Place, Summertown, Oxford OX2 7BX
O and C	Oxford and Cambridge Schools Examination Board 10 Trumpington Street, Cambridge; and Elsfield Way, Oxford
SUJB	Southern Universities' Joint Board for School Examinations Cotham Road, Bristol BS6 6DD
WJEC	Welsh Joint Education Committee 245 Western Avenue, Cardiff CF5 2YX

SCE Board

SEB	Scottish Examination Board Ironmills Road, Dalkeith, Midlothian EH22 1BR

CSE Boards

ALSEB	Associated Lancashire Schools Examining Board 17 Harter Street, Manchester M1 6HL
EAEB	East Anglian Examinations Board The Lindens Lexden Road, Colchester, Essex CO3 3RL
EMREB	East Midland Regional Examinations Board Robins Wood House, Robins Wood Road, Apsley, Nottingham NG8 3NH
LREB	London Regional Examinations Board Lyon House, 104 Wandsworth High Street, London SW18 4LF
NREB	North Regional Examinations Board Wheatfield Road, Westerhope, Newcastle upon Tyne NE5 5JZ
NWREB	North West Regional Examinations Board Orbit House, Albert Street, Eccles, Manchester M30 0WL
SREB	Southern Regional Examinations Board 53 London Road, Southampton SO9 4YL
SEREB	South East Regional Examinations Board Beloe House, 2/4 Mount Ephraim Road, Royal Tunbridge Wells, Kent TN1 1EU
SWEB	South Western Examinations Board 23–29 Marsh Street, Bristol BS1 4BP
WJEC	Welsh Joint Education Committee 245 Western Avenue, Cardiff CF5 2YX
WMEB	West Midlands Examination Board Norfolk House, Smallbrook Queensway, Birmingham B5 4NJ
WY & LREB	West Yorkshire and Lindsey Regional Examining Board Scarsdale House, 136 Derbyshire Lane, Sheffield S8 8SE
YREB	Yorkshire Regional Examinations Board 31–33 Springfield Avenue, Harrogate, North Yorkshire HG1 2HW

Section II Core units 1–28
1 Measurements

All measurements in Physics are related to the three chosen fundamental quantities of **length**, **mass** and **time**. For many years scientists have agreed to use the metric system; the particular one used now is based on the metre, the kilogram and the second.

1.1 LENGTH

The unit of length is the metre. Various multiples or submultiples are also used. Thus:

1 kilometre (km) = 1000 metres;
1 metre   (m) = 100 centimetres = 1000 millimetres;
1 centimetre (cm) = 10 millimetres (mm).

For day-to-day work in laboratories metre and half-metre rules are used – graduated in centimetres and millimetres. For more accurate measurement vernier calipers or a micrometer screw gauge may be used. Details of both these instruments will be found in any standard textbook. A metre rule is accurate to the nearest millimetre, calipers to the nearest 0.1 mm and a micrometer gauge to 0.01 mm.

1.2 MASS

The mass of a body measures the quantity of matter it contains. The unit of mass is the kilogram. A body will have the same mass in all parts of the universe.

$$1 \text{ kilogram (kg)} = 1000 \text{ grams (g)};$$
$$1 \text{ gram} \qquad\quad = 1000 \text{ milligrams (mg)}.$$

1.3 AREA AND VOLUME

The area of a surface is measured in units of metres times metres (m^2) or cm^2 or mm^2. The volume of a substance is usually expressed in units of m^3 or cm^3 or mm^3.

The volume of a liquid is often measured in litres or millilitres.

	1 litre (l)	= 1000 millilitres (ml);
but	1 millilitre	= 1 cm^3 approximately;
therefore	1 litre	= 1000 cm^3 approximately.

In Physics the measuring cylinder is most commonly used for volume measurements of liquids. When reading the value it is important to look at the bottom of the curved liquid surface (meniscus).

1.4 TIME

The scientific unit of time is the second which is $\dfrac{1}{24 \times 60 \times 60}$ part of the time the earth takes to perform one revolution on its axis.

In the laboratory, time is normally measured using a stopclock or stopwatch. In some cases more accuracy is required, as, for example, when measuring the acceleration of a trolley moving on a ramp. A tickertape vibrator can then be used; see unit 2.1 on speed, velocity and acceleration. The use of a centisecond or millisecond timer is explained in unit 2.8. The sensitivities of these four instruments are as follows:

stopwatch or stopclock	1/10 second;
tickertape vibrator	1/50 second;
centisecond timer	1/100 second;
millisecond timer	1/1000 second.

In some experiments it may be appropriate to use a stroboscope to measure time. A description of the hand-operated stroboscope and its use is given at the end of unit 17.2. Similarly the flashing stroboscope and its use are described in units 2.8 and 2.9.

2 Speed, velocity and acceleration

2.1   TICKERTAPE VIBRATOR OR 'TICKER-TIMER'

The motion of an object moving in a laboratory can be studied with the aid of a tickertape, on which equal intervals of time are marked by a dot. The dots are printed by a 'ticker-timer' (Fig. 2.1), which consists of a flexible strip of soft iron A, clamped at one end B, and passing over one pole of a strong electromagnet M. A small alternating voltage from the mains (50 Hz) is connected to the electromagnet through a diode D, which only allows the resulting current to flow in one direction. Fifty times a second the current through the electromagnet grows and then dies away. Each time it grows the iron strip is attracted downwards. A stud or 'hammer' beneath the strip strikes a carbon disc below it, thus marking a tape C running beneath the carbon. Each time the current dies away the strip is released. Thus 50 dots are made every second at equal intervals of $\frac{1}{50}$ second.

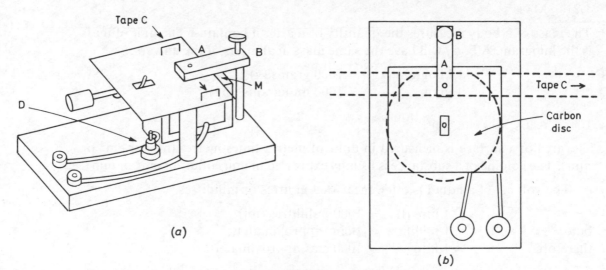

(a)

(b)

Fig. 2.1 (*a*) Ticker-timer; (*b*) plan view

2.2   MOTION

Fig. 2.2 shows tickertape pulled through the timer by a pupil who changed his speed as he went and on one occasion collided with another pupil.

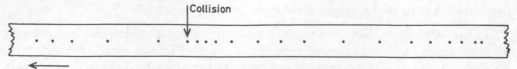

Fig. 2.2 Tickertape studies of motion

To show how the distance s travelled by an object towing a length of tape varies with time t, the tape is cut into consecutive lengths containing equal numbers of dots (*e.g.* five). These lengths represent distances travelled in equal intervals of time (0.1 s), and so measure the speed of the object. If they are arranged side by side (Fig. 2.3) a histogram of speed against time is obtained.

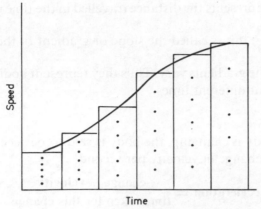

Fig. 2.3 Speed-time histogram and graph

In fact the speed does not change suddenly as indicated by the histogram. Each tape length is a measure of the average speed during $\frac{1}{10}$ second, whereas the speed is really changing all the time. A line drawn through the mid-points of the tape tops is a more accurate indication of the speed of the object at any moment. This line is a graph.

2.3 SPEED

Speed is defined as the distance moved in one second.

$$\textbf{Average speed} = \frac{\textbf{distance moved}}{\textbf{time taken}}\text{(m/s)}$$

Average speed may be obtained by measuring the distance travelled by an object while a certain number of dots (e.g. five) are made on the tape it is towing. For example, the average speed during each five-dot length of tape in Fig. 2.3 is calculated by dividing the length of each tape by 0.1s. It is not, of course, necessary to cut up the tape in order to do this. The average speed can be calculated over any number of dots on any length of tape, providing the correct time interval is used in the calculation. If the dots on a measured length of tape are evenly spaced (*i.e.*, the object is travelling with constant speed), then the speed calculated is the actual speed during that time interval.

2.4 VELOCITY

The velocity of a body measures its speed and the direction in which it travels.

$$\textbf{Velocity} = \frac{\textbf{distance moved in a particular direction}}{\textbf{time taken}}$$

(m/s in a particular direction – north for example).

Uniform velocity means that both the speed and the direction remain constant, as shown in Fig. 2.4(*b*).

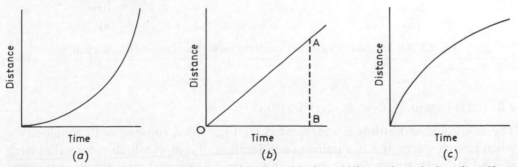

Fig. 2.4 Variation of distance moved in a straight line with time: (*a*) increasing velocity; (*b*) uniform velocity; (*c*) decreasing velocity

In Fig. 2.4(*b*) *AB* represents the distance travelled in the time represented by *OB*,

thus   Velocity $= \dfrac{AB}{OB}$; this is called the slope or gradient of the graph.

In Fig. 2.4(*a*) and (*c*) the gradients vary. Thus they represent bodies whose velocities have different values at different times.

2.5   ACCELERATION

If the velocity of a body is changing, the body is said to be accelerating. Acceleration is defined as the change in velocity per second.

$$\textbf{Acceleration} = \frac{\textbf{change in velocity}}{\textbf{time taken for this change}}$$

For example, suppose a car travelling along a straight road increases its speed from 10 m/s to 20 m/s in five seconds.

Change in velocity $= (20 - 10)$ m/s $= 10$ m/s;
time taken for this change $= 5$ seconds.

Hence using the formula above:

acceleration $= 10$ m/s in 5 seconds $= 2$ m/s in 1 second;
thus acceleration $= 2$ m/s^2.

In this example the acceleration resulted from a change in the magnitude of velocity (speed). However velocity can change in either magnitude or direction (circular motion). A change in either means the body is accelerating.

In Fig. 2.5(*a*) the velocity is increasing with time at a steady rate, and the acceleration is said to be uniform. *PQ* represents the change in velocity in time *OQ*.

$$\text{Acceleration} = \frac{PQ}{OQ} \text{ the gradient of the velocity-time graph.}$$

In Fig. 2.5(*b*) the acceleration is not constant. The acceleration at any instant is found by calculating the gradient of the graph at that time.

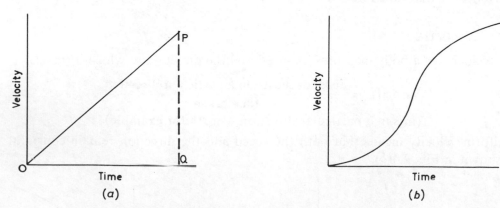

Fig. 2.5 Acceleration: (*a*) uniform acceleration; (*b*) non-uniform acceleration

2.6   UNIFORMLY ACCELERATED MOTION

The body whose motion is represented by Fig. 2.6 is moving with a velocity *u* when timing starts. It has a **uniform acceleration** of *a* m/s^2 which means that each second its velocity increases by *a*, and after *t* (seconds) its velocity will have increased by *at*. Hence at the end of this time its velocity $v = u + at$.

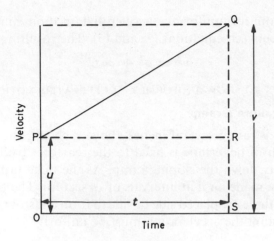

Fig. 2.6

Alternatively,

$$\text{acceleration } a = \frac{\text{change in velocity}}{\text{time taken for this change}}$$

$$= \frac{v - u}{t};$$

therefore $\qquad\qquad v - u = at$

and $\qquad\qquad v = u + at.$ $\qquad\qquad$ (1)

2.7 DISTANCE TRAVELLED

The average velocity of the body whose motion is shown in Fig. 2.6 is equal to half the sum of its initial velocity u, and final velocity v.

Average velocity $= \dfrac{u + v}{2}$ (this only applies to a body accelerating uniformly).

The distance s moved can be found using the equation:

$$\text{velocity} = \frac{\text{distance moved}}{\text{time taken}}$$

hence $\qquad$ distance moved $=$ average velocity $\times$ time taken.

$$s = \frac{(u + v)}{2} \times t \qquad\qquad (2)$$

but $\qquad\qquad v = u + at.$ $\qquad\qquad$ (1)

Therefore $\qquad\qquad s = \dfrac{(u + u + at)t}{2}$

or $\qquad\qquad \boldsymbol{s = ut + \tfrac{1}{2}at^2}.$ $\qquad\qquad$ (3)

In Fig. 2.6, ut is the area of the rectangle $OPRS$. The area of the triangle

$$PQR = \tfrac{1}{2} \times \text{base} \times \text{height}$$
$$= \tfrac{1}{2} \times t \times (v - u),$$

but $\qquad\qquad v - u = at$ from equation (1).

Thus $\qquad\qquad$ area $PQR = \tfrac{1}{2}at^2,$

therefore the area of $OPQS =$ (area of rectangle $OPRS$) + (area of triangle PQR).

Hence $\qquad\qquad$ area of $OPQS = ut + \tfrac{1}{2}at^2$

but $\qquad\qquad s = ut + \tfrac{1}{2}at^2.$ $\qquad\qquad$ (3)

The area under the velocity–time graph in Fig. 2.6 is therefore equal to the distance travelled by the body. This is true for all such graphs, even when the acceleration is non-uniform.

A third equation for uniformly accelerated motion may be obtained by eliminating time t between equations (*1*) and (*3*). The resulting equation is

$$v^2 = u^2 + 2as. \qquad (4)$$

2.8 EXPERIMENTS TO SHOW UNIFORMLY ACCELERATED MOTION

1. A trolley moving down a ramp

One end of a ramp is raised several inches above a bench; the other end rests on the bench. A length of tickertape is fixed to the rear of a trolley, and the trolley allowed to run freely down the sloping ramp. As the tickertape passes through a ticker-timer dots are made on it at intervals of $\frac{1}{50}$ second. The tape is then cut into five dot lengths and these are placed side by side to form a histogram (Fig. 2.7). Since the gradient is constant the acceleration must be uniform.

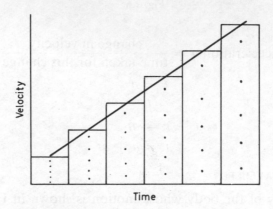

Fig. 2.7 Uniform accelerated motion

2. A body falling freely under gravity

In the last experiment the acceleration of the trolley was fairly small. A body falling freely under gravity has a much larger acceleration, and therefore the time of fall which has to be measured is small. It is usual to use a centisecond or millisecond timer for this purpose.

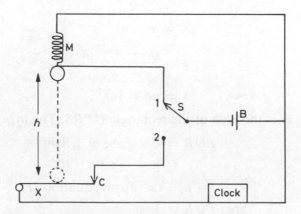

Fig. 2.8 Free fall method for *g*

The apparatus used for this experiment is shown in Fig. 2.8. It consists of an electromagnet M connected to a battery B via a switch S. The current passing through the electromagnet provides a magnetic field which holds a metal ball in place. When the switch is moved to position 2, the current stops and the ball begins falling. At the same time the circuit to the clock (centisecond or millisecond timer) is completed and it starts counting. After falling through a height h, the ball hits a hinged plate X, thus breaking the contact at C and stopping the clock. The time t is read to an accuracy of $\frac{1}{100}$ or $\frac{1}{1000}$ second. The distance h is measured as accurately as possible; the value of the acceleration of gravity is then calculated using equation (3).

$$s = ut + \tfrac{1}{2}at^2.$$

Here
$$s = h, u = 0, a = g,$$

thus
$$h = \tfrac{1}{2}gt^2,$$

hence
$$g = \frac{2h}{t^2}.$$

It is usual to repeat the experiment once or more as a check. The value of the acceleration due to gravity is approximately 9.8 m/s^2 at any point on the earth's surface. It has a much lower value on the surface of the moon.

An alternative way of timing the fall of the ball which does not use the electromagnet or the hinged plate is shown in Fig. 2.9. The ball is released by hand such

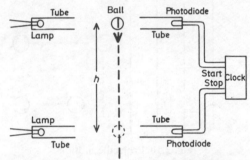

Fig. 2.9 Free fall method for g using photodiodes

that it falls between each lamp and the photodiode opposite it. When the ball is between each lamp and its photodiode it prevents light from the lamp reaching the photodiode. The photodiodes are connected to the clock (centisecond or millisecond timer) in such a way that when the ball passes the upper diode the clock is made to start and when the ball passes the lower diode the clock is stopped. The distance h between the two lamps or photodiodes is measured with a rule and the acceleration due to gravity g is calculated in the same way as above.

In a third method for doing this experiment a stroboscope is used. It may be either a hand-operated (unit 17.2) or a flashing one. A flashing stroboscope emits short flashes of light at regular intervals. The time between each flash can be changed by means of a calibrated dial. For this experiment a suitable time interval between successive flashes is 0.05s (20 flashes per second).

The stroboscope is arranged so that the flashes from it illuminate a white ball as it falls in a dark room. A millimetre scale is placed alongside the path of the falling ball. As the ball is released the shutter of a camera is opened and this is closed when the ball stops falling. The film in the camera thus records a series of images of the ball alongside the scale at known equal time intervals. The distance h the ball falls during a known time t is measured from the picture and the acceleration due to gravity g calculated as in the previous methods.

In each of these alternative methods measurements can be made over different heights h. Whatever value of h is used the value of g is about the same; that is, g is found to be constant.

2.9   PROJECTILES

So far we have only considered objects travelling in a straight line. In this unit we shall study objects which are being accelerated by the force due to gravity and at the same time are moving horizontally at a steady speed. One example of such motion is the path of a ball which is projected horizontally over the edge of a table. This ball continues to move with the same horizontal speed as it had just as it left the edge of the table (if we ignore the small effect of air resistance). In addition the ball falls under the influence of gravity.

It can be shown that the ball takes the same time to reach the ground as another ball dropped from the same height at the moment the first ball leaves the edge of the table. This shows that the horizontal motion of the first ball in no way affects its vertical acceleration; that is, it falls in exactly the same way as it would if it were not moving sideways. The horizontal and vertical motions of the ball can be treated entirely separately.

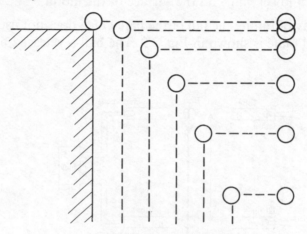

Fig. 2.10 Projectile motion

Fig. 2.10 represents the motions of the two balls already mentioned. The paths of the two have been photographed at regular intervals. It can be seen that in equal time intervals both balls fall an increasing but equal distance (they undergo the same vertical acceleration). In the same equal time intervals the first ball moves equal distances horizontally. This motion may also be demonstrated using the pulsed water drops experiment. Another example of projectile motion is that of a ball thrown from one person to another.

---

3 Force, momentum, work, energy and power

3.1   NEWTON'S LAWS OF MOTION

The majority of the work discussed in this unit can be summarised in Newton's Laws of Motion:

1. **Every body continues in a state of rest or uniform motion in a straight line unless acted on by a force.**
2. **When a force acts on a body the rate of change of momentum of the body is proportional to the force and the momentum changes in the direction in which the force acts.**
3. **To every action there is an equal and opposite reaction.**

The word **force** denotes a push or pull. When a body at rest is acted on by a force it tends to move; if a force acts on a body already in motion it will change its velocity, either by altering its speed or its direction, or both. If a body has no force acting on it, the body will remain at rest or it will continue to move with a steady velocity.

Force is that which changes, or tends to change, a body's state of rest or uniform motion in a straight line.

A spring balance marked in newtons is suitable for measuring force. The balance consists of a spring whose extension is proportional to the force applied to it. The spring is contained in a case, and is calibrated by applying known forces to it, usually in the form of weights (Fig. 3.1).

3.2 THE ACCELERATION PRODUCED BY A FORCE

The relationship between a force and the acceleration it produces can be investigated using a trolley, ramp and ticker-timer. The apparatus is the same as the one described in the last unit for showing uniformly accelerated motion. The ramp has to be raised sufficiently at one end to compensate for friction in the trolley and ticker-timer. The slope should be such that when the trolley with tape in place, is nudged, it moves slowly down the ramp at constant speed.

A tape is then obtained by towing the trolley down the ramp with a constant force. This can be provided by either an elastic band stretched a known amount,

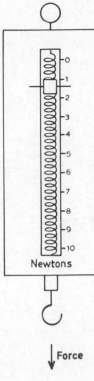

Force

Fig. 3.1

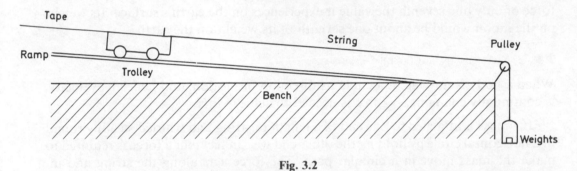

Fig. 3.2

or by weights hung over a pulley as shown in Fig. 3.2. The experiment is repeated using twice the original force and again using three times the force.

Histograms are made from the tapes in the usual way. A typical set of results is shown in Fig. 3.3.

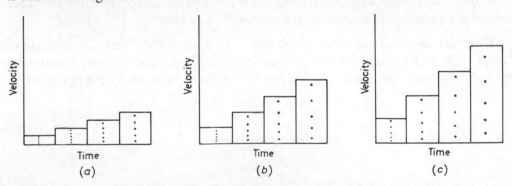

Fig. 3.3 (*a*) Original force; (*b*) twice the force; (*c*) three times the force

It is clear that if the force doubles, the acceleration doubles *etc*., that is

$$\frac{F}{a} \text{ is constant, provided the mass is constant.} \tag{5}$$

The same apparatus may be used to show how the acceleration (*a*) depends on the mass (*m*) of a trolley, when a constant force is used. The force is applied in turn to one, two and three trolleys of similar mass stacked on each other. It is found that if the mass is doubled the acceleration is halved *etc.,*

that is, $m \times a$ is constant.   (6)

The relation $F = ma$ (7)

will be seen to include both statements (5) and (6). The unit of force is the **newton (N)** which is defined as the force which gives an acceleration of one metre per second² to a mass of one kilogram.

An object which has a large mass requires a large force to accelerate it as can be noted from equation (7). Such an object is said to have large inertia; it is diffiicult to move.

3.3   WEIGHT

The acceleration produced by the earth when a body is falling freely is written as *g*. The force on a body due to the earth's attraction is found by using equation (7) which becomes $F = mg$, where F is in newtons.

mg is the value of the force acting on a body due to the earth's attraction, or its weight. This force acts downwards, that is towards the centre of the earth.

The strength of the moon's attraction is about one seventh of the value of the earth's attraction. A mass *m* on the moon's surface would therefore experience a force of only one seventh the value it experiences on the earth's surface. Its weight on the moon would be about one seventh of its weight on the earth.

3.4   MOTION IN A CIRCLE

When a body is moving in a circle its direction of motion, and hence its velocity, is continually changing. A force is needed to achieve this.

If a mass is attached to one end of a length of string and whirled steadily round in a horizontal circle by holding the other end we can feel that a force is required to make the mass move in a circular path. The force acts along the string and increases if the stone is whirled faster.

If any object is to move in a circle a force has to be continually applied to the object at right angles to its direction of motion at any instant; that is towards the centre of the circle. This is called the centripetal force. It does not alter the speed of the object, but it does alter its direction of travel and hence its velocity. The force produces an acceleration towards the centre of the circle.

Consider an object of mass *m* moving with a constant speed *v* in a circle of radius *r*. By calculating the rate of change of *velocity*, as the object continually changes direction, the acceleration *a* can be shown to be given by the equation:

$$a = \frac{v^2}{r},$$

but $F = ma$

thus $F = \frac{mv^2}{r}.$ (8)

The relationship shown in equation (8) can be investigated using the apparatus illustrated in Fig. 3.4. It consists of a mass *S*, such as a rubber stopper, at the end of a long piece of string or cord *C*, which passes through a glass tube with a rubber grip. The ends of the glass tube are polished so that friction is reduced and there is little danger of the string fraying. The force *F* is provided by a number of steel washers *W* of equal weight attached to the end of the string.

The mass *S* is swung in a horizontal circle at a steady speed, with as little movement of the hand as possible. Care is taken that the paper clip rotates freely just

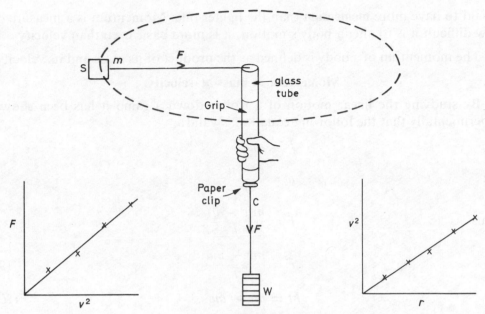

Fig. 3.4 Centripetal force

below the grip. This ensures that the radius remains constant. The time is noted for a given number of revolutions: 20 for example.

Then $v = \dfrac{2\pi rn}{T}$ where n is the number of revolutions counted and T is the time noted.

The number of washers is then increased and the experiment repeated using the same radius. A series of results is obtained in this way. A further series is obtained using a constant force, but changing the radius.

The graphs shown in Fig. 3.4. are plotted. They show that:

(i) $\dfrac{F}{v^2}$ is constant for a constant radius,

(ii) $\dfrac{v^2}{r}$ is constant for a constant force.

The equation $\boldsymbol{F = \dfrac{mv^2}{r}}$ includes both these statements.

If the string had broken at any time the mass S would have flown off at a tangent to the circle; that is, on the removal of the force, the mass would have continued in a straight line with a steady speed.

A car or train going round a corner are everyday examples of motion in a circle. Friction between the tyres and the road provides the centripetal force in the case of a car cornering. If the corner is too sharp (r very small), the speed too high, or the road wet, the frictional force may not be large enough to keep the car moving in a circle, and it will slide off the road. The outer rail provides the centripetal force in the case of a train.

Another example of centripetal force is the gravitational force of the sun on the planets, including the earth.

3.5   MOMENTUM

A lorry which is fully laden requires a larger force to set it in motion than a similar lorry which is empty. Likewise more powerful brakes are required to stop a heavy goods vehicle than a family car moving with the same speed. The heavier vehicle is said to have more **momentum** than the lighter one. Momentum is a measure of how difficult it is to alter a body's motion; it is more basic even than velocity.

The momentum of a body is defined as the product of its mass and its velocity.

$$\textbf{Momentum = mass} \times \textbf{velocity.}$$

By studying the linear motion of a trolley down a ramp it has been shown experimentally that the following equation is valid:

$$F = ma$$

but

$$a = \frac{v - u}{t}.$$

Thus

$$F = \frac{m(v - u)}{t},$$

so

$$F = \frac{mv - mu}{t} \tag{9}$$

or

$$Ft = mv - mu. \tag{10}$$

Equation (9) can be written thus:

$$\textbf{Force} = \frac{\textbf{change of momentum}}{\textbf{time taken for this change}}.$$

All the equations just discussed concern momentum change and hence, by the definition of momentum, velocity change. They are therefore valid for changes in the direction of motion of a body as well as for changes in its speed. They apply to circular motion as well as linear, and are summed up in Newton's second law of motion.

3.6   ACTION AND REACTION

When two bodies collide the force one exerts on the other is equal in size but opposite in direction to the force the second one exerts on the first. As the time of contact is the same for each both experience the same change in momentum, but in the opposite direction. The total momentum of the two is thus unaltered by the collision. This is known as **the principle of conservation of linear momentum**. The principle always holds providing no forces, other than those due to the collision, act on the bodies.

This principle can be verified experimentally, for a simple case, using two trolleys and a ticker-timer, on a friction-compensated ramp. Fig. 3.5(a) shows the arrangement.

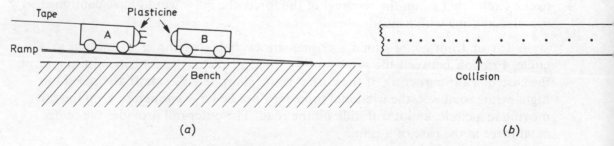

Fig. 3.5 Momentum

Some plasticine is placed on the surface of the two trolleys which will come into contact. Drawing pins are embedded in the plasticine on the front of trolley A, and tickertape is attached to the rear of this trolley. Trolley A is given a push so that it travels with a uniform velocity and collides with trolley B. The trolleys stick together and move off with a common velocity. A typical tape is shown in

Fig. 3.5(*b*). The spacing of the dots changes abruptly on collision. The velocity of *A* before the collision and the combined velocity of the two trolleys after the collision are found from the spacing of the dots. In the experiment the tape is **NOT** being used to show acceleration. The results show that:

Total momentum before collision = total momentum after collision.   (*11*)

Alternatively

$$m_1 u_1 + m_2 u_2 = m_1 v_1 + m_2 v_2 \tag{12}$$

where *m*, *u* and *v* represent the masses and velocities of the colliding bodies. In some collisions one of the velocities may be zero or two velocities may have the same value. For example in the experiment just described $u_2 = 0$ and $v_1 = v_2$, as the bodies stick together after collision. Equation (*12*) then becomes simpler.

Suppose that in the above experiment $m_1 = 1$ kg, $m_2 = 3$ kg and $u_1 = 1$ m/s. Then substituting in equation (*12*) we obtain:

$$1 = v_1 + 3v_2$$

but $$v_1 = v_2,$$

thus $$1 = 4v_1,$$

hence $$v_1 = 0.25 \text{ m/s} = v_2.$$

This is the value we would record from the tape after the collision, thus verifying **the principle of the conservation of momentum**.

The principle of conservation of momentum holds in the case of an explosion as well as a collision. A rocket relies for its propulsion on the fact that it gains a forward momentum equal in size to the momentum of the gases it expels backwards. It will gain forward momentum, and hence increase speed, for as long as it continues to expel these gases.

The principle may be used to estimate the speed of a pellet fired from an air rifle. The experiment can be done in the laboratory using the apparatus shown in Fig. 3.6. The pellet is fired into a piece of plasticine embedded in a block of wood mounted on a suitable vehicle. The vehicle is stationary on a track which has been compensated for friction. The speed of the vehicle after the collision is found by timing its motion over a measured length of track with a stopwatch. The mass of one pellet is determined by weighing 10 or 20 similar pellets.

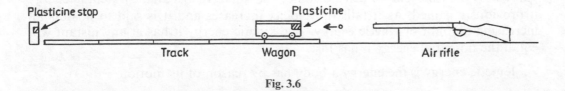

Fig. 3.6

The initial speed of the pellet is calculated using equation (*12*) above. Again $u_2 = 0$ and $v_1 = v_2$. All the other quantities in the equation can be measured, except u_1, which is calculated from it. The value is likely to be about 100 m/s.

3.7   WORK

The term 'work' is associated with movement. If a railway engine pulls a train along a track with a steady force the work done by the engine depends on the size of the force it provides and the distance it pulls the train with this force. Work is calculated by definition from the relation:

Work done = force × distance moved in the direction of the force.

The unit of work is the joule (J). One joule is the work done when a force of one newton moves through a distance of one metre in the direction of the force.

No work is done in circular motion as the force is at right angles to the distance.

3.8   ENERGY

Anything which is able to do work is said to possess energy.

Energy is the capacity to do work. All forms of energy are measured in joules.

The world we live in provides energy in many different forms, of which chemical energy is perhaps the most important. The use of chemical energy from coal and oil has been a major factor in the development of our civilisation. The presence of electricity, light and heat as forms of energy in our homes is something we take for granted. All these are produced from the chemical energy released from coal or oil or from nuclear energy.

Energy can neither be created nor destroyed, though it may be changed from one form to another. This is a statement of **the law of conservation of energy**.

Although energy may change from one form to another, the second form may not be measurable or useful. For example this is true of the heat produced when a frictional force is overcome.

3.9   POTENTIAL ENERGY

If a body is to be raised from a bench, an upward vertical force equal to the weight of the body must be provided. Suppose the force applied (*mg*) raises the body a vertical distance *h*.

Work done = force × distance moved in the direction of the force.

Thus work done = $mg \times h =$ **mgh.**

Once the body has been raised it is said to have increased its **potential energy** by this amount. All the work done has been used to increase the potential energy of the body.

Potential energy is the energy a body has by reason of its position.

3.10   KINETIC ENERGY

If the mass is now allowed to fall it will steadily lose the potential energy it has gained. By the time it has fallen a distance *h* it will have lost all the potential energy it previously gained. As it falls its velocity increases and it is said to possess an increasing amount of **kinetic energy**. The kinetic energy it has at any instant will equal the potential energy it has lost.

Kinetic energy is the energy a body has by reason of its motion.

As the mass falls it is accelerated by the force of the earth's attraction on it. Suppose it falls a distance *h*, in time *t*, as its velocity increases uniformly from zero to *v*.

$$\text{Force} = \frac{\text{change in momentum}}{\text{time taken}}.$$

Thus
$$F = \frac{mv - 0}{t}$$

and   distance = average velocity × time,

that is
$$h = \frac{v}{2} \times t.$$

Work done = force × distance moved in the direction of force

$$= \frac{mv}{t} \times h$$

$$= \frac{mv}{t} \times \frac{vt}{2}$$

$$= \tfrac{1}{2}mv^2.$$

The work done by the gravitational force results in an increase of kinetic energy of $\frac{1}{2}mv^2$, and a loss of potential energy of *mgh*.

Suppose a body, mass 5 kg, falls through a vertical distance of 5 m near the earth's surface where its acceleration due to gravity (*g*) is 10 m/s².

The potential energy lost $= mgh = 5 \times 10 \times 5 = 250$ J.

The kinetic energy gained $= \frac{1}{2}mv^2 = 250$ J.

Thus $\qquad\qquad\qquad\qquad \frac{1}{2} \times 5 \times v^2 = 250$ J,

$$v^2 = 100$$

and $\qquad\qquad\qquad\qquad\qquad v = 10$ m/s.

3.11 POWER

Power is defined as the work done per second, or the amount of energy transformed per second.

$$\textbf{Average power} = \frac{\textbf{work done}}{\textbf{time taken}} = \frac{\textbf{energy change}}{\textbf{time taken}}.$$

It is measured in units of joules per second (J/s). One joule per second is called a watt (W).

$$1 \text{ kilowatt (kW)} = 1000 \text{ watts.}$$

A rough estimate of a pupil's power can be made by asking him to walk or run up a staircase. If the height of the staircase is measured in metres, the pupil's weight calculated in newtons, and the time he takes recorded on a stopwatch, the power can be calculated. The result will normally be about 200 watts if walking or 500 watts if running.

4 The parallelogram and triangle of forces

4.1 SCALAR AND VECTOR QUANTITIES

A **scalar quantity** is one which has only magnitude (size), such as money and number of apples. A **vector quantity** is one which has both magnitude and direction, such as velocity, force and momentum.

Vectors can be represented by straight lines drawn to scale. If a number of forces all act in the same straight line, their resultant is determined by addition or subtraction.

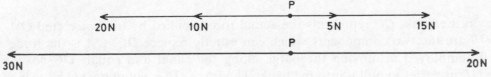

Fig. 4.1

Thus if forces of 20, 15, 10 and 5 newtons all act at a point *P*, as shown in Fig. 4.1, we have:

total force acting towards the left = 10 + 20 = 30 N;

total force acting towards the right = 15 + 5 = 20 N;

resultant force = 30 − 20 = 10 N acting to the left.

If two forces act at a point, but not in the same straight line, the principle of the parallelogram of forces may be used to find their resultant.

4.2   The parallelogram of forces

If two forces acting at a point are represented both in magnitude and direction by the adjacent sides of a parallelogram, their resultant is represented in both magnitude and direction by the diagonal of the parallelogram drawn from the point.

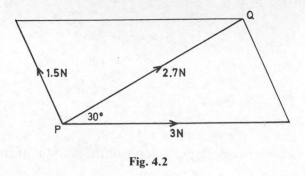

Fig. 4.2

Fig. 4.2 shows how this principle is applied. The resultant of 2.7 N is given by the length of the diagonal *PQ*. Its direction makes an angle of 30° with the force of 3 N.

4.3   Resolution of forces

The principle of the parallelogram of forces has just been used to find the resultant of two forces acting at a point. To find the effectiveness of a force in a given direction the reverse process is used and the force is 'resolved' into two components at right angles to each other.

Fig. 4.3 shows the forces involved when a barge is towed along a canal by a

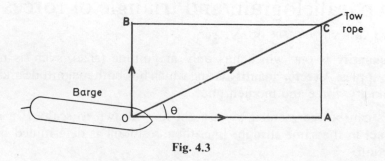

Fig. 4.3

horse on the bank. *OC* represents the actual force applied by the horse, and *OA* and *OB* are the two components which can exactly replace *OC*. *OA* is the force usefully employed in moving the barge along the canal and equals $OC \cos \theta$, while *OB* just tries to pull it into the bank. The force *OB* is counteracted by using the rudder to point the barge slightly outwards.

4.4   THE TRIANGLE OF FORCES

If three forces acting at a point are in equilibrium, they can be represented in magnitude and direction by the three sides of a triangle taken in order. 'Taken in order' means that the arrows showing the force directions must follow each other in the same direction round the triangle.

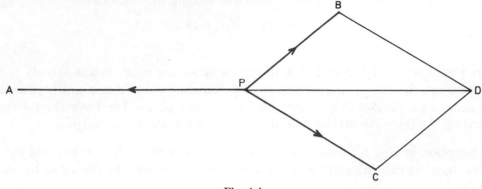

Fig. 4.4

Consider the three forces *PA*, *PB*, *PC* acting at the point *P* in Fig. 4.4. The resultant of *PB* and *PC* is represented by *PD*. If the three forces are in equilibrium *PD* must be equal in magnitude and opposite in direction to *PA*. As *PBDC* is a parallelogram *PC = BD* and they are parallel. Thus the three forces *PA*, *PB*, *PC* can be represented in magnitude and direction by *DP*, *PB* and *BD*, forming a closed triangle.

---

5 Turning forces and machines

When we open a door, turn on a tap or use a spanner, we exert a **turning force**. Two factors determine the size of the turning effect: the magnitude of the force and the distance of the line of action of the force from the pivot or fulcrum. A large turning effect can be produced with a small force provided the distance from the fulcrum is large. The size of the turning effect is called the **moment**.

The moment of a force about a point is the product of the force and the perpendicular distance of its line of action from the point.

Moment of force = force × perpendicular distance from the pivot.

5.1   EXPERIMENT TO STUDY MOMENTS

A thin uniform strip of wood, for example, a metre rule, is balanced on a fulcrum. A weight is placed on the rule to one side of the fulcrum, a second weight added

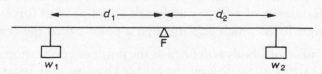

Fig. 5.1 Moments

on the other side and its position carefully adjusted until balance is restored (Fig. 5.1).

Within the limits of experimental error it will be found that:

$$w_1 \times d_1 = w_2 \times d_2.$$

Balance can be restored using more than one weight, in which case:

$$w_1 \times d_1 = (w_2 \times d_2) + (w_3 \times d_3).$$

This principle can be extended if further weights are used. When a body is in equilibrium, the sum of the anti-clockwise moments about any point is equal to the sum of the clockwise moments about the same point. The force the fulcrum exerts on the body equals the sum of the turning forces on the body.

Suppose, in Fig. 5.1 above, $w_1 = 300$ N, $w_2 = 100$ N, $d_1 = 0.4$ m and $d_2 = 0.5$ m, then we can calculate where a weight of 700 N must be placed to balance the rule.

$$w_1 \times d_1 = (w_2 \times d_2) + (w_3 \times d_3),$$

hence

$$300 \times 0.4 = (100 \times 0.5) + 700d_3$$

and

$$d_3 = \frac{(300 \times 0.4) - (100 \times 0.5)}{700} \text{ m}$$

$$= \frac{120 - 50}{700} = 0.1 \text{ m}.$$

This weight must be placed 0.1 m to the right of the pivot.

5.2   PARALLEL FORCES

Parallel forces acting in the same direction are called **like forces**, and it is always possible to find their resultant or a single force which exactly replaces them.

Parallel forces acting in opposite directions are called **unlike forces**. In most cases they can be replaced exactly by a single force, but this is not so in the case of two equal and opposite (unlike) parallel forces. They form what is called a couple, which causes rotation and can only be balanced by an equal and opposite couple (Fig. 5.2).

Fig. 5.2 A couple

5.3   CENTRE OF MASS

The weight of a body is defined as the force with which the earth attracts it. This says nothing about the point of application of this force. A body may be regarded as made up of a large number of tiny particles, each with the same mass. Each of these particles is pulled towards the earth with the same force. The earth's pull on the body thus consists of a large number of equal parallel forces. These can be replaced by a single force which acts through a point called the **centre of mass**.

The centre of mass of a body is defined as the point of application of the resultant force due to the earth's attraction on the body. Thus we may regard the centre of mass as the point at which the whole weight of the body acts.

5.4 LOCATION OF THE CENTRE OF MASS

The centre of mass of a long thin object such as a ruler may be found approximately by balancing it on a straight edge. The same method may also be used for a thin sheet (lamina). In this case it is necessary to balance it in two positions as shown in Fig. 5.3.

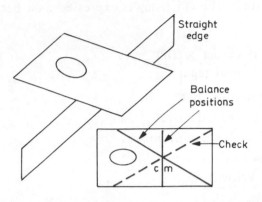

Fig. 5.3 Locating the centre of mass   **Fig. 5.4**

One good way of finding the centre of mass of a lamina is to use a plumbline. Three small holes are made at well spaced intervals round the edge of the lamina, and the lamina and plumbline suspended from each in turn (Fig. 5.4).

The position of the plumbline is marked on the lamina and the point of intersection of these three lines gives the position of the centre of mass.

5.5 STABILITY

The position of a body's centre of mass affects its stability. For example, the centre of mass of a vehicle should be as low as possible, and its wheel base as wide as possible, if it is to be stable.

When a vehicle corners fast there is a tendency for it to tilt on the outer wheels. It will turn right over when the vertical line through the centre of mass falls outside these wheels. If the conditions stated above are satisfied the centre of mass has to rise a larger distance for this to happen. This requires more potential energy and it is less likely to occur.

5.6 MACHINES

A machine is any device by means of which a force (effort) applied at one point can be used to overcome a force (load) at some other point. Most machines, but not all, are designed so that the effort is less than the load. They can be regarded as force multipliers; the force that the effort exerts is multiplied by the machine to be equal to the force that the load exerts.

The **mechanical advantage** of any machine is defined as the ratio of the load to the effort.

$$\text{Mechanical advantage} = \frac{\text{load}}{\text{effort}}.$$

For example, if a lever can be used to overcome a load of 500 N by the application of an effort of 100 N, it has a mechanical advantage of five.

The ratio of the distance moved by the effort to the distance moved by the load in the same time is called the **velocity ratio** of the machine.

$$\text{Velocity ratio} = \frac{\textbf{distance moved by the effort}}{\textbf{distance moved by the load in the same time}}.$$

Both mechanical advantage and velocity ratio are dimensionless; that is, they have no units.

5.7   WORK DONE BY A MACHINE: EFFICIENCY

The ratio of the useful work done by a machine to the total work put into it is called the **efficiency** of the machine. Usually the efficiency is expressed as a percentage.

$$\textbf{Efficiency} = \frac{\textbf{work output} \times \textbf{100\%}}{\textbf{work input}}.$$

Since work done = force × distance, it follows that:

$$\text{efficiency} = \frac{\text{load} \times \text{distance load moves} \times 100\%}{\text{effort} \times \text{distance effort moves}}$$

thus

$$\text{efficiency} = \frac{\text{mechanical advantage} \times 100\%}{\text{velocity ratio}}$$

In a perfect machine no work would be wasted and the efficiency would be 100%. It follows that the mechanical advantage and velocity ratio are then equal. In practice, work is wasted in overcoming friction and, in the case of pulleys, in raising the lower pulley block. The efficiency is then below 100%.

5.8   THE LEVER

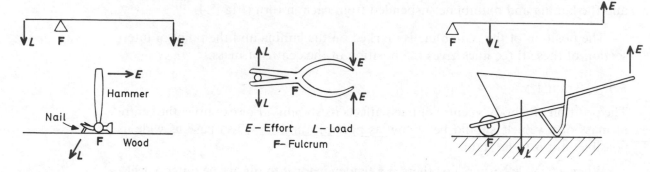

Fig. 5.5 Simple levers

The lever is the simplest form of machine in common use. It can consist of any rigid body pivoted about a fulcrum. Levers are based on the principle of moments discussed earlier in this section. A force (effort) is applied at one point on the lever, and this overcomes a force called the load at some other point. Fig. 5.5 illustrates some simple machines based on the lever principle.

Details of other machines such as the screw jack, inclined plane and gears may be found by reference to standard textbooks.

5.9   PULLEYS

A pulley is a wheel with a grooved rim. Often one or more pulleys are mounted together to form a block. The use of pulleys is best illustrated by considering the block and tackle arrangement shown in Fig. 5.6.

For clarity the pulleys are shown on separate axles; in practice the two pulleys in the top block are mounted on a single axle as are those in the bottom block. In order to raise the load by one metre each of the four strings supporting the lower block must be shortened by 1 m. This is achieved by the effort being applied through a distance of 4 m.

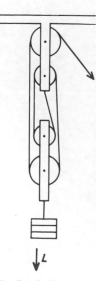

Fig. 5.6 Pulley system

6 Density

Equal volumes of different substances vary considerably in their mass. For instance aircraft are made chiefly from aluminium alloys which, volume for volume, have a mass half that of steel, but are just as strong. The 'lightness' or 'heaviness' of a material is referred to as **its density**.

$$\text{Density} = \frac{\text{mass}}{\text{volume}} \text{ kg/m}^3 \text{ or g/cm}^3.$$

6.1 DENSITY MEASUREMENTS

A regular solid

The mass of the solid is found by weighing, either on a chemical balance, if great accuracy is required, or on a spring balance. If the latter is used the result must be converted to mass units; that is if the balance is calibrated in newtons, the value recorded must be divided by 9.81 to obtain the mass in kilograms.

The volume of the solid is obtained by length measurement, using a ruler, vernier calipers or a micrometer screw gauge, depending on the accuracy required. This method is applicable to cuboids, spheres, cylinders and cones amongst other regular shapes. The formula giving the volume of such shapes in terms of their linear dimensions can be obtained from textbooks. For example, metals are often in the form of turned cylinders whose volume can be calculated from the formula:

Volume $= \pi r^2 h$ where r is the cylinder radius and h is its height.

An irregular solid

The mass is found in the same way as for a regular solid. In order to find the volume it is necessary to partly fill a measuring cylinder with water. The reading is taken and the solid then lowered into the water on the end of a length of cotton, until it is completely immersed, and the new reading taken. The difference between the two readings gives the volume of the solid. This method cannot be used if the solid dissolves in water.

A liquid

A measuring cylinder is first weighed empty using, for example, a top-pan balance. Some of the liquid to be tested is poured into the cylinder, and the cylinder re-weighed. The difference between the two readings gives the mass. The volume of the liquid is obtained by direct reading of the measuring cylinder. If a more accurate value is required a specific gravity bottle can be used. Details may be obtained from a textbook.

A gas

Air is the most common gas. Its density may be found by pumping air into a large plastic container. The container, complete with an open tap, is first weighed; a lever arm balance is suitable for this purpose. Air is then pumped into the container until it is very hard, using a bicycle or foot pump and a one-way valve in the connecting tube. The tap is then closed and the container re-weighed. By means of tubing attached to the tap, the air is now released from the container and collected over water as shown in Fig. 6.1.

A transparent plastic box B makes a suitable collecting container. A line L is marked on it so that one litre is collected. The box is filled with water and inverted as shown. Air slowly bubbles through the tube, and the box is moved up or down until the outside and inside water levels are in line with the mark L. One litre of air at atmospheric pressure has now been collected. The tap is closed, the plastic

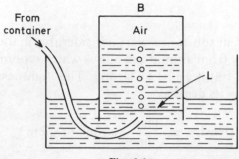

Fig. 6.1

box refilled with water and the bubbling process repeated. This is continued until no more air bubbles through. The final fraction of a litre collected has to be estimated.

The mass and volume of the extra air originally pumped into the container are now known and the density of air at atmospheric pressure may be calculated.

6.2   RELATIVE DENSITY

The heaviness of a substance is often expressed by comparing its mass or weight with the mass or weight of an equal volume of water. The ratio is called **the relative density**.

$$\text{Relative density} = \frac{\text{mass of any volume of a substance}}{\text{mass of an equal volume of water}}.$$

7   Pressure in liquids and gases

The word **pressure** has a precise scientific meaning. It is defined as the force acting normally (perpendicularly) per unit area. For example, the pressure exerted on the ground by a body depends on the area of the body in contact with the ground. A boy wearing ice skates will exert a far greater pressure than a boy wearing shoes. The pressure exerted on the ground by a brick depends on which face is in contact with the ground. The weight of the brick and thus the force it exerts on the ground is about 22 N, whichever face is in contact (Fig. 7.1).

$$\text{Pressure} = \frac{\text{force}}{\text{area}}.$$

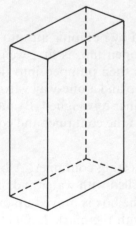
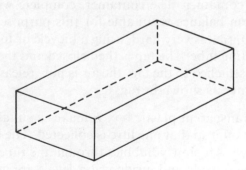

Fig. 7.1

One does not have to use a very large force when using a needle. As the area of the point is very small, a relatively small force produces a large pressure and the needle pierces the material.

7.1 PRESSURE IN A LIQUID OR GAS (A FLUID)

The pressure in a fluid increases with depth. This may be shown by using a tall vessel full of water with side tubes fitted at various depths (Fig. 7.2).

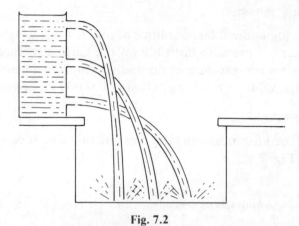

Fig. 7.2

The speed with which the water spurts out is greatest for the lowest jet, showing that pressure increases with depth. This demonstration also indicates that pressure acts in all directions in a fluid – not just vertically. The pressure responsible for these jets is acting horizontally.

Suppose we consider a horizontal area A at a depth h below the surface of a fluid of density d (see Fig. 8.2). Standing on this area is a vertical column of liquid of volume hA, the mass of which is hAd. The weight of this mass is $hAdg$.

$$\text{Pressure} = \frac{\text{force (weight)}}{\text{area}} = \frac{hAdg}{A}$$

$$= hdg$$

The usual units are newtons/metre^2, often called pascals (Pa). The area does not appear in the final expression for pressure.

The property of liquids to transmit pressure to all parts is used in many appliances. Some car jacks consist of an oil-filled press used for lifting (Fig. 7.3).

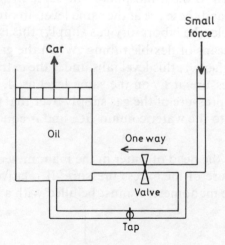

Fig. 7.3 Hydraulic car jack

Mechanical diggers and bulldozers use hydraulic principles to power the blade or shovel. Cars require a braking system which exerts the same pressure on the brake pads of all four wheels to reduce the risk of skidding. Each brake consists of two brake shoes which are pushed apart by hydraulic pressure in a cylinder and press on the brake drum. When the brakes are applied the increased pressure on the pedal is transmitted through oil to the cylinders in each wheel and the shoes applied.

7.2   ATMOSPHERIC PRESSURE

On earth we are living under a large volume of air. Air has weight and as a result the atmosphere exerts a pressure not only on the earth's surface but on objects on the earth. **Atmospheric pressure** is normally expressed in newtons/metre^2 or pascals. The average value over a long period is 100 000 Pa approximately.

7.3   THE MANOMETER

This instrument is used for measuring the pressure of a gas. It consists of a U-tube containing water (Fig. 7.4).

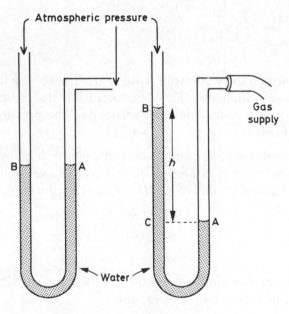

Fig. 7.4

When both arms are open to the atmosphere, the same pressure is exerted on both water surfaces *A* and *B* and these are at the same level. In order to measure the pressure of a gas (for example, the laboratory gas supply) this is connected to one arm of the manometer by means of flexible tubing. When the gas supply is turned on it exerts a pressure on surface *A*; this level falls under the extra pressure and the level at *B* rises, until the pressure at *C*, on the same level as *A*, becomes equal to the gas pressure. The excess pressure of the gas supply over that of the atmosphere is given by the pressure due to the water column *BC*, and is equal to *hdg* (newtons per square metre).

The height *h* is known as the head of water in the manometer and it is common to use this height as a measure of the excess pressure. If relatively high pressures are to be measured using the manometer it must be filled with a more dense liquid such as mercury.

7.4   SIMPLE BAROMETER

The simple barometer uses the same principle as the manometer; the reservoir takes the place of one arm. As atmospheric pressure is to be measured it is necessary to have a vacuum above the mercury in the tube (Fig. 7.5).

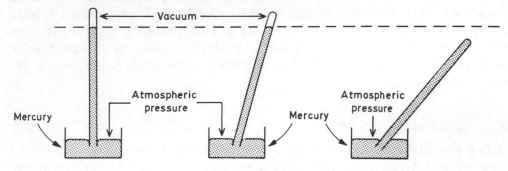

Fig. 7.5 Simple barometer

A simple barometer may be made by taking a thick-walled glass tube about one metre long and closed at one end, and filling it almost to the top with mercury. A finger is then placed securely over the end and the tube inverted several times. The large air bubble which travels up and down collects the small ones clinging to the side of the tube. More mercury is then added so that the tube is completely full. A finger is again placed over the open end, the tube inverted carefully, its open end placed under the surface of mercury in a reservoir or dish, and the finger removed. The mercury falls until the vertical difference in level between the two mercury surfaces is about 76 cm. Small changes in the level of the mercury in the tube occur from day to day due to small changes in atmospheric pressure.

7.5   ANEROID BAROMETER

An aneroid barometer contains no liquid. The basis of the instrument is a flat cylindrical metal box which has been partially evacuated and sealed (Fig. 7.6).

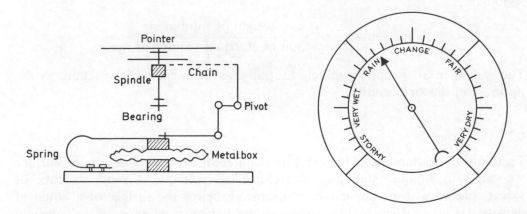

Fig. 7.6 Aneroid barometer   **Fig. 7.7**

Increase in atmospheric pressure causes the box to cave in slightly; a decrease allows it to expand slightly. These movements of the box are magnified by a system of levers. The levers are connected to a fine chain wound round the spindle of a pointer. A hairspring keeps the chain taut. The pointer moves over a dial usually calibrated as shown in Fig. 7.7.

The dial of an aneroid barometer can be calibrated in feet for use as an altimeter in aircraft. Atmospheric pressure falls by about $\frac{1}{3}$ during the first 10 000 feet of ascent above the earth's surface.

8 Buoyancy

When any object is placed in a liquid or gas it experiences an upward force called the **upthrust**. This effect may be illustrated by a very simple experiment. A brick immersed in water may be lifted by a piece of cotton fixed to it. However, immediately one attempts to lift the brick by this means in air, the cotton breaks. The brick experiences a relatively large upthrust while under water; in air this force is much smaller.

8.1 ARCHIMEDES' PRINCIPLE

The first experiments to measure the upthrust of a liquid were carried out by the Greek scientist Archimedes in the third century B.C. He established the principle which states:

When a body is wholly or partially immersed in a fluid (liquid or gas) it experiences an upthrust equal to the weight of fluid displaced.

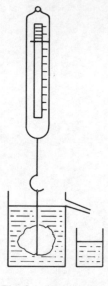

This principle may be verified in the following way. A displacement can is positioned on the bench with a beaker under its spout. The can is filled with water until it overflows and then the beaker is replaced by another which has previously been dried and weighed. A suitable solid such as a piece of metal or stone is suspended from a spring balance and its weight in air noted. The solid is then very carefully lowered until it is completely under the water in the can, the excess water being collected in the beaker (Fig. 8.1). The weight of the submerged solid is noted, and the beaker containing the water weighed. The difference in value between the weighings of the solid is found to be the same as the difference in value between the weighings of the beaker. The apparent loss in weight of the body has been shown to equal the weight of water displaced.

Fig. 8.1 Archimedes' principle

The set of readings recorded can be used to determine the relative density of the solid as:

$$\text{relative density} = \frac{\text{weight of solid in air}}{\text{weight of an equal volume of water}}.$$

The weight in air is approximately the same as the weight in a vacuum as the upthrust of the air is small.

8.2 THEORY

From our understanding of the pressure at different depths below a liquid surface we can gain an understanding of Archimedes' principle. Consider a cube of side x, placed with its upper face a distance h below the surface of a liquid of density d (Fig. 8.2). The hydrostatic pressure on the vertical faces of the cube will

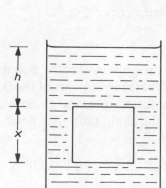

Fig. 8.2

cancel as corresponding points on opposite faces will experience an equal but opposite pressure.

$$\text{Pressure} = hdg,$$

thus the total downward force on the top face of the cube will be

$$hdgx^2$$

and that acting vertically upwards on the bottom face

$$(h + x)dgx^2.$$

The resultant upward force on the cube is:

$$(h + x)dgx^2 - hdgx^2 = dgx^3.$$

This is the weight of liquid displaced by the cube.

Although more difficult to show, the principle applies to solids of all shapes.

8.3   BALLOONS

A hydrogen-filled balloon rises when released in air for precisely the same reason that a cork rises when placed below water. The upthrust is greater than the weight of the hydrogen and the fabric of the balloon, as hydrogen is much less dense than air.

8.4   SHIPS

If a body needs only to be partially immersed in a liquid for the upthrust to equal its weight in air, then it will float on the surface. The law of flotation is a special case of Archimedes' principle. It states:

A floating body displaces its own weight of the fluid in which it floats.

Because a ship is hollow and contains air its average density even when laden is less than that of water.

8.5   HYDROMETERS

A body will float at different levels in liquids of different density. This is the principle of the hydrometer (Fig. 8.3).

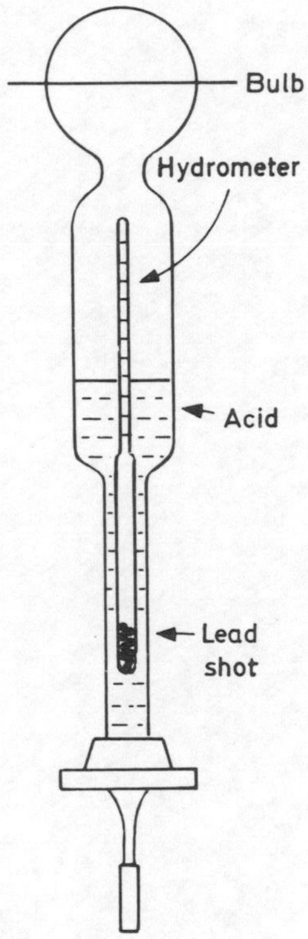

The lower bulb is weighted with lead shot or mercury to keep it upright and the upper stem graduated to read the density or relative density of liquids. The stem is thin so that the instrument is more sensitive (the marks are well spread). Fig. 8.3 shows a type used for measuring the relative density of accumulator acid. The lower end of the hydrometer is immersed in the acid, the bulb squeezed to expel air, and then released, whereupon acid is pushed into it by the atmospheric pressure acting on the acid surface. The density or relative density is then read on the floating hydrometer.

Fig. 8.3 Hydrometer

9 Forces between molecules in solids

Before materials are used in the construction of machinery, bridges and buildings, tests are carried out to ensure that they are able to stand up to the stresses to which they are likely to be subjected. Brittle substances such as cast iron and masonry will support large forces of compression but break easily if stretching forces are applied. When stretching forces are likely to be significant, materials such as steel have to be used. The behaviour of a material under the influence of applied forces depends on the forces holding the molecules of the material together.

9.1 ELASTICITY

Some knowledge of the forces between molecules of a solid can be gained by adding weights to a spiral spring and investigating how it stretches. A spiral spring is suspended vertically from a rigid support and a small pointer attached to its lower end (Fig. 9.1(*a*)). The reading of the pointer against the scale is noted. Weights are then added in steps to the lower end of the spring and the reading of the pointer recorded after the addition of each weight. The weights are removed in similar steps and a second set of readings taken. For small loads the reading of the pointer should be the same for each set of readings. The average extension for each load is then plotted against the load (Fig. 9.1(*b*)).

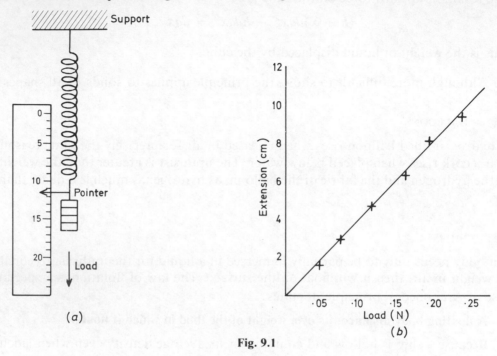

Fig. 9.1

A straight-line graph passing through the origin shows that the extension is directly proportional to the load in the range of loads used: that is, the load/extension ratio is constant. When a small weight is attached to the spring, the spring extends. When the weight is removed the spring returns to its original length. The property of regaining its original size or shape is called **elasticity** – thus putty is a very inelastic material whereas metals regain their original shape and are therefore elastic. In a metal the forces of attraction between the displaced molecules are sufficiently strong to restore the molecules to their original position.

If larger weights had been added to the spring a stage would have been reached when the spring would not have returned to its original shape on removing the weights. The spring is permanently stretched or deformed. Beyond a certain load the molecules do not return to their original positions when the load is removed. The extension at which this occurs is called the **elastic limit** of the spring. With greater loads the molecules are unable to keep their fixed positions in the metal.

The forces between the molecules of a metal can be further investigated by stretching a length of straight wire. A length of wire is suspended vertically from a fixed support and weights added to its lower end, in a similar way to the procedure for a spring. However, the extension of the wire within its elastic region is very small, and if this region is to be studied a more accurate method of measuring extension has to be used. It is usual to use two wires, one carrying a vernier and the other a millimetre scale (Fig. 9.2(*a*)). The second wire is for comparison purposes and is not stretched. This comparison method of measuring extension eliminates errors due to thermal expansion and sag of the support.

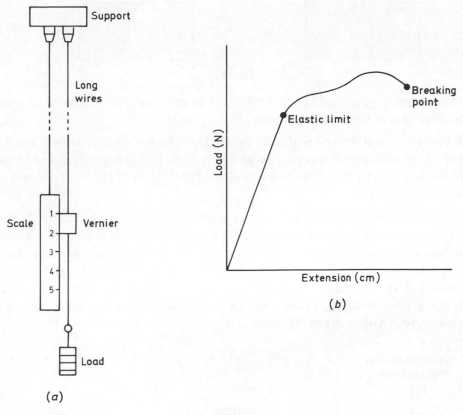

Fig. 9.2

The results from these two experiments may be summarised in what is known as Hooke's law:

The deformation of a material is proportional to the force applied to it, provided the elastic limit is not exceeded.

10 Forces between molecules in liquids

A sewing needle can be made to float on water, although its density is considerably greater than that of water. The needle can be placed on a small piece of filter paper, which is then gently placed on the water surface. Within a few seconds the paper sinks to the bottom and the needle floats.

Close examination of the water surface reveals that the needle rests in a slight depression. The surface of water always tends to become as small as possible and resists forces tending to make it larger. This property of a liquid is called **surface tension**. If the needle sank, it would first need to increase the depression of the surface, and hence increase the area. The surface tension of the water, in resisting this increase, prevents the needle from sinking. If a few drops of alcohol, soap solution or detergent are added to the water, the surface tension is decreased and the needle sinks.

The tension in the surface of a liquid is well illustrated by a soap film. A wire frame with a piece of cotton tied across it is dipped into soap or detergent solution so that a film is formed. When the film on one side of the cotton is broken by touching it, the tension in the film on the opposite side pulls the cotton into the arc of a circle (Fig. 10.1).

When an irregularly shaped piece of camphor is placed on the surface of water, it dissolves most readily at its points. The surface tension of the solution is less

Fig. 10.1

than that of pure water and hence the piece of camphor moves on the water surface in the direction of the resultant force.

A tent keeps out water because of the surface tension which exists in the lower surface of the rain water in contact with the material. It is important not to touch the inside of the material or the water will wet the strands of the material and soak through at this point.

The tension which exists in the water surface or soap film is due to the forces which exist between molecules of the liquid concerned. Inside the liquid a molecule is surrounded equally on all sides by neighbouring molecules. On the surface, however, a molecule has many more molecules below it than above it in the vapour. Consequently there tends to be a resultant force on the surface molecule pulling it towards the interior of the liquid. This makes the liquid have the minimum surface area for a given volume (Fig. 10.2(a)).

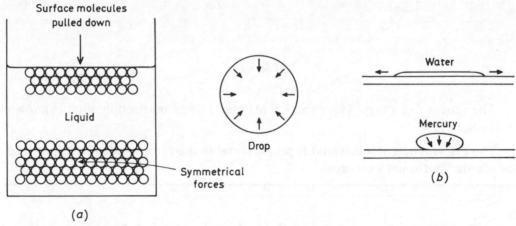

Fig. 10.2

Water dripping from a tap is seen to take a spherical shape when falling. This is due to the forces between water molecules just described. However, when water is spilled on a clean glass surface it wets the glass and spreads out in a thin film. The force between water molecules and glass molecules is greater than between water molecules. However, the opposite is true in the case of mercury and glass. Thus, when mercury is spilled on glass it forms small spherical drops (Fig. 10.2(b)).

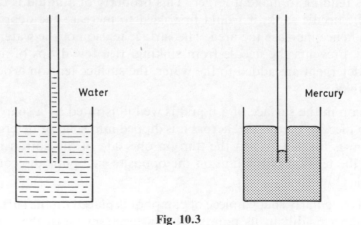

Fig. 10.3

This difference between the molecular properties of water and mercury explains why the meniscus of water curves upwards while the meniscus of mercury curves downwards when those liquids are poured into clean glass vessels.

If a length of glass tubing with a very fine bore (a capillary tube) is dipped into water, or any other liquid which wets glass, it is noticed that the liquid rises in the tube to a height of several centimetres. However, if the same procedure is followed with mercury a depression results (Fig. 10.3).

In the case of liquids that wet glass the forces between the liquid molecules are less than the forces between the liquid and glass molecules. The meniscus curves upwards and the liquid column is being supported by the upward force of surface tension between the glass and liquid acting round the circumference of the meniscus.

11 Kinetic theory of matter

11.1 ATOMS AND MOLECULES

In 1808 John Dalton produced experimental evidence to show that chemical compounds consist of molecules. A molecule is a group of atoms. There are about 90 different chemical elements occurring in nature each of which has its own characteristic atom.

Some idea of the size of molecules can be gained from an experiment to measure the thickness of a very thin oil film. A tank with a large surface area is required for this experiment and the water which is placed in the tank must be very clean. Lycopodium or talcum powder is lightly sprinkled on the water surface. A small oil drop is then placed on the surface of the water near the centre of the tank. It can be transferred to the water using a loop of fine wire or a small syringe. Immediately the oil drop touches the water surface the powder is pushed back by the oil film to form a ring of clear water, the diameter of which is measured.

The oil film may be considered as a cylinder of radius R and height h. Hence its volume $= \pi R^2 h$. The value of h is calculated by equating the volume of the oil film to the volume of the oil drop placed on the water surface. On extremely clean water, olive oil is considered to spread until the film is one molecule thick. In this case h is an estimate of the diameter of an oil molecule. The value obtained is about 2×10^{-9} m.

11.2 THE STATES OF MATTER

The molecules in a solid are each anchored to one position, about which they vibrate continuously, as if held in position by a framework of springs. When heat energy is supplied, thus raising the temperature, the molecules vibrate faster and through greater distances than before. Thus the extra energy is transformed into kinetic energy of the molecules.

If sufficient heat energy is supplied a solid will melt to form a liquid. The amplitude of vibration of the molecules becomes so large that they break away from the position to which they were anchored and move freely amongst each other. However, forces still act between the molecules holding them close to each other and so although a liquid has no definite shape it does have a definite volume. The volume of a liquid is much the same as the solid from which it forms, thus the average separations of the molecules are about the same in each case, as are the densities. The molecular separation in the liquid may be slightly greater than in the solid (*e.g.* wax) or slightly less (*e.g.* water).

Not all the molecules have exactly the same energy at a particular temperature. Some molecules in a liquid have more than average energy and if they are near the liquid surface they are able to escape; this is evaporation. If heat energy is supplied to the liquid the average energy of each molecule increases; that is they move faster. Eventually all the molecules have sufficient energy to break away from each other and so a gas is formed, in which all molecules move independently. This change of state is known as boiling.

The molecules of a gas move continually, colliding with each other and with the walls of their containing vessel. The laws of mechanics apply to these collisions; further they are elastic, that is, no kinetic energy is lost. When a molecule bounces back from the walls of the container its momentum is changed and so it must have experienced a force. The force the molecules exert on the walls accounts for the pressure. If a gas is heated its molecules move faster and exert a greater pressure on the walls. An example of this effect is that if the pressure of a car or bicycle tyre is measured on a hot day it will be found to be greater than on a cold day, even though no gas has been allowed to enter or leave the tyre in the meantime. In a gas the forces between the molecules are so small that the molecules can be considered to move independently of each other. A gas therefore fills all the available space and has no fixed volume or shape; it takes the volume and shape of the vessel containing it. When the pressure exerted by a gas is equal to atmospheric pressure its volume will be about 1000 times greater than the liquid from which it forms, and its density therefore about 1000 times less. This can be shown using the syringes illustrated in Fig. 11.1.

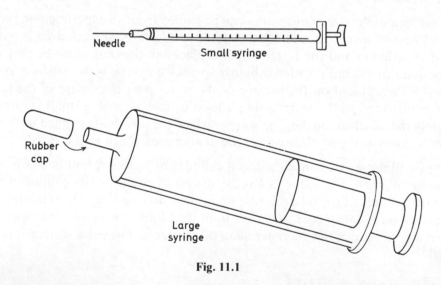

Needle
Small syringe
Rubber cap
Large syringe

Fig. 11.1

A rubber cap is fitted over the end of the large syringe with its piston pushed down to zero volume. The small syringe with a hypodermic needle at its end is partially filled with water. 0.1 ml of water is then injected into the large syringe through the rubber cap via the hypodermic needle. When the needle is removed the cap seals. The large syringe is now inverted in a beaker of brine and the brine brought to the boil. The water in the syringe turns to steam and is seen to occupy a volume of about 100 ml; that is 1000 times larger than when it was water.

11.3 BROWNIAN MOTION

The continual motion of molecules within a liquid or gas is called **Brownian motion**, after Robert Brown. In 1827 he used a microscope to examine pollen particles sprinkled on the surface of water and was surprised to notice that they were in a continuous state of haphazard movement. It appears that the motion of these relatively large particles is caused by the impact of moving water molecules.

The same kind of movement can be seen in the case of smoke particles in air. The apparatus for this experiment is shown in Fig. 11.2.

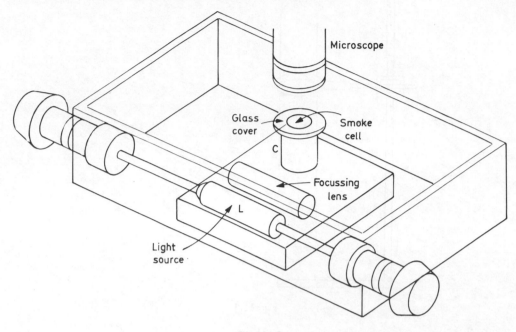

Fig. 11.2

It consists of a small transparent cell *C*, with a cover, strongly illuminated from the side by a light *L*. A piece of cord or rag is set smouldering and some of the smoke which contains minute particles is collected by a syringe and injected into the cell. The cover is replaced and the microscope focused on the cell. The particles are seen to be moving in an irregular way, darting about suddenly, and always in motion.

The irregular motion of a particle is due to the movement of air molecules, which bombard it from all sides. The particle is relatively small and so the number of air molecules hitting it on one side is not balanced by an equal number hitting the opposite side at the same instant. The smoke particle thus moves in the direction of the resultant force. The irregular motion of the particles shows that air molecules move rapidly in all directions. At higher temperatures the molecules move even more rapidly and the motion of the particles is even more violent and irregular.

If the particles in a gas are very much bigger than the molecules of the gas they do not show this irregular movement. This is because the large number of molecules hitting one side is not relatively much greater than the number hitting the other at the same moment. The resultant force is thus relatively very small and the large particle is not so easily moved. For example, a table tennis ball suspended in air does not move for this reason.

11.4 DIFFUSION

Diffusion provides further evidence for the irregular random motion of molecules. Consider the apparatus shown in Fig. 11.3.

The vertical tube *T* initially contains air. A capsule containing liquid bromine is placed in the end tube *X* and this connected to the tap *Y* by means of a short length of thick-walled rubber tubing. The tap is closed and the capsule tapped until it moves down inside the rubber tubing. The tubing is then squeezed with a pair of pliers until the capsule breaks. Liquid bromine runs down to the tap which is

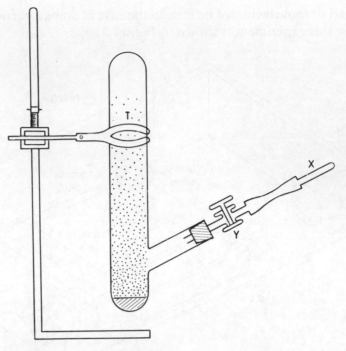

Fig. 11.3

then opened allowing the bromine to enter the tube *T*. If the tube is observed some minutes later the characteristic brown colour of the bromine will be seen to have diffused part way up the tube. This is made more obvious by placing a white card behind the tube. If the experiment is now repeated, with the tube *T* initially evacuated of air, the brown bromine vapour will fill the entire tube the moment the tap is opened.

The behaviour of the bromine in this experiment can only be explained by saying that the molecules of the bromine gas continually collide with air molecules in the first case. Their progress up the tube is thus impeded, whereas in the second case there are no air molecules to get in their way.

11.5   RANDOMNESS AND MEAN FREE PATH

The continual collisions of molecules within a gas means that the path of an individual molecule is as shown in Fig. 11.4.

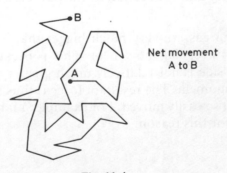

Net movement
A to B

Fig. 11.4

Although the actual speed of a molecule is very high (about 500 m/s or 1000 miles per hour) the net movement in one direction is very small because it continually doubles back on its path. For this reason, if the stopper is removed from a bottle containing a pungent smelling liquid or gas, the smell will take some minutes to reach the far side of the room.

The average distance travelled by a molecule between collisions is termed **the mean free path**. Because of the **randomness** of collisions a molecule will travel different distances between collisions, but the average may be calculated (about 10^{-7} m at atmospheric pressure). The mean free path increases as the density of a gas is reduced.

11.6 SUMMARY

All matter is composed of molecules in continuous motion. In a solid they vibrate about a fixed position. In a liquid they move freely amongst each other, whereas in a gas they are entirely independent of each other. The small size of molecules is demonstrated by the oil drop experiment, and their continuous movement by Brownian motion. The pressure exerted by a gas is accounted for by the momentum change when its molecules hit the containing walls.

12 Expansion of solids and liquids

12.1 SOLIDS

As temperature increases, the molecules of a solid vibrate with greater amplitude, thus causing the total volume of the solid to increase; that is, a solid expands when heated. This expansion can be troublesome and very large forces may be set up if there is an obstruction to the free movement of the expanding body. Continuous welded rails used on most of our railway lines are held in place by strong concrete clamps and sleepers. They are stress free at 25°C but experience very large forces at high and low temperatures.

Modern motorways are often constructed with a concrete surface for economy. If this surface were laid in one continuous section cracks would appear owing to the expansion and contraction brought about by the differing summer and winter temperatures. To avoid this, the surface is laid in short sections, each section being separated from the next by a small gap filled with black pitch. On a hot day the pitch is squeezed out by the expansion of the concrete. Allowance has to be made for the expansion of bridges and the roofs of buildings made of steel girders. A common way of overcoming these difficulties is to fix one end of the structure while the other rests on steel rollers.

Although expansion can be troublesome it can be useful. Steel plates such as those used in ship building are often rivetted together using red hot rivets. Holes are made in the overlapping plates, a red hot rivet pushed through and its head pressed tightly against one plate. The other end of the rivet is hammered until it is tight against the second plate. As the rivet cools it contracts thus pulling the two plates together. A watertight join is formed.

When strips of the same length but of different substances are heated through the same range of temperature, their expansions are not always equal. This difference may be used to make a thermostat. Fig. 12.1 shows the principle of a thermostat which is a device for maintaining a steady temperature.

The heater circuit is completed through the two contacts at *C*. One of the contacts is attached to the end of a metal strip *S*, the other to the end of a bimetallic strip *M*. On heating the brass expands more than the invar and the strip bends with the brass on the outside. At a certain temperature the strip bends so much that the contacts are pulled apart thus breaking the circuit to the heater. When the air cools the bimetallic strip straightens, the contacts close and the heater switches on again. The thermostat may be set to operate at different temperatures by

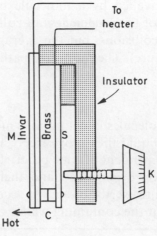

Fig. 12.1

adjusting the knob *K*. If the knob pushes the metal strip *S* towards the bimetallic strip a higher temperature is maintained.

Thermostats working on the same principle are used to control the temperature of electric irons, immersion heaters, aquaria for fish and for other purposes.

12.2   EXPANSIVITY

The coefficient of linear expansivity of a substance is the fraction of its original length by which a rod of the substance expands per degree rise in temperature.

If α is the expansivity of a material then each metre length of it expands by α metres for each degree rise in temperature.

$$\text{Coefficient of linear expansivity} = \frac{\text{increase in length}}{\text{original length} \times \text{rise in temperature}}.$$

The coefficient may be measured using the apparatus shown in Fig. 12.2. The length of the metal rod, about 0.5 m, is measured with a metre rule. The rod is

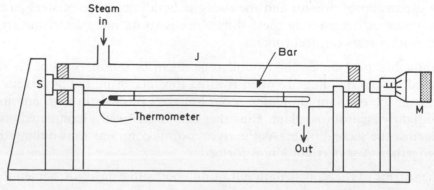

Fig. 12.2

then enclosed in a steam jacket between a fixed stop *S* and a micrometer *M*. The micrometer is closed until the rod is held firmly between it and the stop, and the micrometer reading noted. The thermometer is also read. The micrometer is then unscrewed several turns, to allow for the expansion of the rod, and steam from a boiler passed through the jacket *J* for several minutes. The micrometer is again tightened, its reading taken, and also that of the thermometer. The fractional increase in length of the rod is now calculated and divided by the temperature rise to give the expansivity.

12.3 LIQUIDS

Liquids expand more than solids on heating. This may be shown by means of a flask fitted with a rubber bung and a length of glass tubing as shown in Fig. 12.3

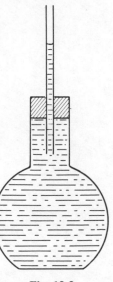

The flask is filled with a liquid and the bung pushed in until the level of liquid comes a short distance up the tube. When the flask is plunged into hot water the liquid level is first seen to fall and then to rise. The initial fall is due to the expansion of the glass which becomes heated first and expands before the heat has had time to reach the liquid. Due to the expansion of the glass, the expansion of the liquid measured is less than its true expansion.

If we take some water at 0°C and begin to heat it the water contracts instead of expanding over the temperature range 0°–4°C. Its behaviour is unusual. At about 4°C the water reaches its smallest volume, and thus its greatest density. Above 4°C water expands in the normal way. The unusual behaviour of water means that between 0° and 4°C water is less dense the colder it is. The colder water rises to the surface of a pond and it is here that ice first forms. A sheet of surface ice acts as a heat insulator and it takes a long time for the water below to freeze, thus thickening the ice. It is most unlikely that a pond of reasonable depth will freeze right through; thus pond life is able to survive below the ice.

Fig. 12.3

12.4 THERMOMETERS

We must be careful to distinguish the temperature of a body from the internal energy it contains. Temperature is a measure of the degree of hotness of a body whereas the energy it contains depends on its nature and mass as well as its temperature.

The temperature of a substance is a number which expresses its degree of hotness on some chosen scale. It is measured by means of a thermometer. Some thermometers depend on the expansion of a liquid when heated, some on the expansion of a bimetallic strip, and others on the change of other physical quantities brought about by heating; for example electrical resistance.

The most common thermometer in use is that which relies on the expansion of mercury in a glass tube when the mercury is heated. The mercury is contained in a bulb at the lower end of the tube. Above the mercury is some nitrogen. The glass tube is calibrated by dividing it into 100 equal divisions between two fixed points. The upper fixed point is marked on the tube when the thermometer is surrounded by steam boiling under standard atmospheric pressure. The lower fixed point is marked when the thermometer is in pure melting ice. The upper and lower fixed points are given the numbers 100 and zero respectively. This procedure establishes the Centigrade or Celsius scale of temperature.

Mercury freezes at −39°C and therefore a mercury in glass thermometer is not suitable for use in countries such as Russia and Canada which have very cold winters. In such places alcohol in glass thermometers are used, as alcohol remains liquid down to −115°C. However, alcohol boils at a little above 70°C and so is not suitable for high temperatures.

A clinical thermometer is one specially designed to measure the temperature of the human body. It is only necessary for it to have a range of a few degrees on

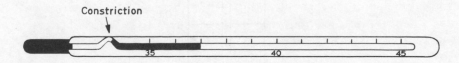

Fig. 12.4

either side of normal body temperature (37°C). The thermometer is generally placed beneath the patient's tongue and left there for two minutes to ensure it acquires the body temperature. The stem has a narrow constriction in its bore just above the bulb (Fig. 12.4). Thus when the thermometer is removed from the mouth, the mercury beyond the constriction stays put while that below it contracts into the bulb. When the temperature reading has been taken, the mercury in the tube is returned to the bulb by shaking.

13 The behaviour of gases

The volume of a gas can be changed not only by altering its temperature, but also by changing the pressure exerted on it. Thus a gas has three quantities: **volume**, **temperature** and **pressure** all of which may change. In order to make a full study of the behaviour of a fixed mass of gas three separate experiments are therefore carried out to investigate:

1. the relation between volume and pressure at constant temperature (Boyle's law);
2. the relation between volume and temperature at constant pressure (Charles' law);
3. the relation between pressure and temperature at constant volume (pressure law).

13.1   Boyle's law

The volume of a fixed mass of gas is inversely proportional to the pressure, provided the temperature remains constant; that is, the pressure multiplied by the volume is constant.

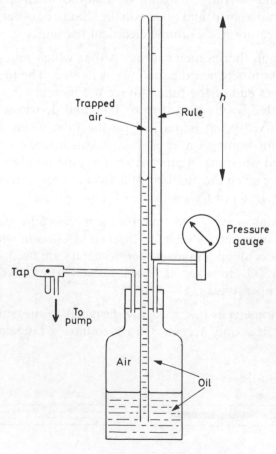

Fig. 13.1

$pV = $ constant

One version of the apparatus used to show this law is illustrated in Fig. 13.1. It consists of a column of air trapped in a vertical tube by some oil with a low vapour pressure. Pressure is applied to the oil in the reservoir by a pump. The Bourdon gauge measures the pressure of the air above the oil in the reservoir. This is a little greater than the pressure of the air trapped in the tube, due to the vertical oil column, but the error is so small that for practical purposes it may be ignored.

Air is first pumped into the reservoir until the Bourdon gauge reaches its maximum reading. The tap is closed and readings taken of the length h of the trapped air column and also the pressure reading of the Bourdon gauge p. The tap is then opened to allow a little air to escape, closed again, and a further set of readings recorded. This procedure is repeated until the Bourdon gauge registers atmospheric pressure once more. It is possible, using a suction pump, to obtain readings below atmospheric pressure. If h is now plotted against $1/p$ a graph is obtained similar to that in Fig. 13.2.

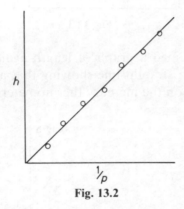

Fig. 13.2

As the volume (V) is proportional to the length of the column of trapped air (h), the fact that the graph is a straight line through the origin shows that:

$$V \div 1/p = \text{a constant}$$

or

$$pV = \text{a constant}$$

or

$$p_1 V_1 = p_2 V_2 \text{ etc.} \tag{1}$$

13.2   CHARLES' LAW

The volume of a fixed mass of gas is directly proportional to its absolute temperature provided the pressure remains constant; that is, the volume divided by the absolute temperature is constant.

$$\frac{V}{T} = \text{constant}$$

A simple apparatus for verifying this law is shown in Fig. 13.3. It consists of a capillary tube sealed at its lower end. Some air has been trapped in the tube by a short thread of mercury M. A centimetre scale is attached to the tube so that the length of the trapped air column can be easily noted. The capillary tube and scale are placed in a beaker of water alongside a thermometer T.

The temperature of the water and the length of the air column are first noted. The water is then heated through about 10°C, time allowed for the heat to reach the air, and the temperature and length of the air column recorded. This process is repeated several times. The volume V of trapped air is proportional to the length of the column as the tube is of uniform bore. The trapped air is kept at constant pressure by the mercury index moving up the tube as the temperature increases.

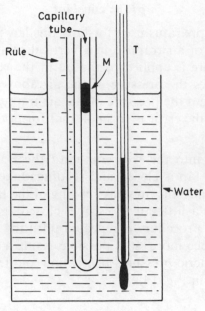

Fig. 13.3

From the results obtained a graph of length against temperature is plotted (Fig. 13.4). The graph is a straight line showing that air expands uniformly with temperature as measured on the mercury thermometer.

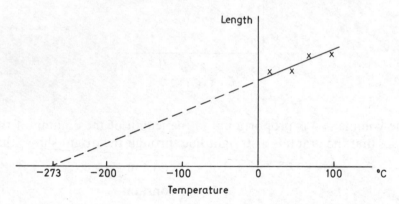

Fig. 13.4 Variation of volume with temperature at constant pressure

If the graph is produced backwards it cuts the temperature axis at a point which gives the temperature ($-273°C$) at which the volume of the gas would contract to zero, assuming the gas continues to contract uniformly below $0°C$. As we cannot imagine it possible for a gas to have a volume of less than zero, it is reasonable to assume that $-273°C$ is the lowest temperature it is possible to obtain, and thus represents the absolute zero of temperature. This assumption cannot be directly tested by experiment as gases liquefy before they reach this temperature, and so the gas laws no longer apply. However, experiments have shown that while temperatures close to $-273°C$ have been reached, it has not been possible to go below this value.

The value $-273°C$ has thus been taken as the zero of a new scale of temperature called the **Absolute** or **Kelvin** scale. Temperatures on this scale are represented by T, and are expressed in units of K. Temperatures on the Celsius scale are converted to the Kelvin scale by adding 273. Thus $0°C = 273$ K.

The graph in Fig. 13.4 is a straight line through the origin of our new temperature scale. Thus the volume of the gas is proportional to its temperature measured on the Kelvin scale and we may write:

$$V \propto T$$

or $$\frac{V}{T} = \text{a constant} \qquad (2)$$

or $$\frac{V_1}{T_1} = \frac{V_2}{T_2} \; etc.$$

13.3 PRESSURE LAW

The pressure of a fixed mass of gas is directly proportional to its absolute temperature provided its volume remains constant; that is, the pressure divided by the absolute temperature is constant.

$$\frac{p}{T} = \text{constant}$$

The apparatus for demonstrating this law is shown in Fig. 13.5. It consists of a

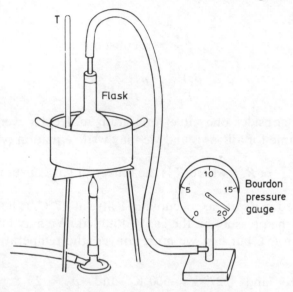

Fig. 13.5

flask connected by rubber tubing to a Bourdon gauge. The flask is surrounded by water in a beaker. A thermometer T is also used.

The apparatus, particularly the rubber tubing, is first inspected for leaks. The temperature of the water and the reading on the Bourdon gauge are noted. The water is then heated while being stirred. After the temperature of the water has risen by about 10°C, heating is stopped, time allowed for the heat to reach the air in the flask, and then the temperature and the pressure are again noted. This procedure is repeated several times until the water is near boiling.

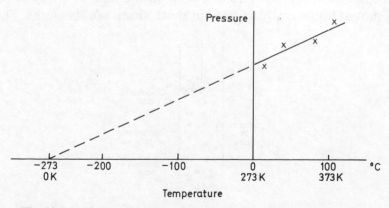

Fig. 13.6 Variation of pressure with temperature at constant volume

From the results a graph of pressure against temperature is plotted (Fig. 13.6). It is a straight line passing through −273°C or 0 K. Thus the pressure of the gas is proportional to its temperature measured on the Kelvin scale and we may write:

$$p \propto T$$

or

$$\frac{p}{T} = \text{a constant} \tag{3}$$

or

$$\frac{p_1}{T_1} = \frac{p_2}{T_2} \ etc.$$

This apparatus can be used to measure temperature (constant volume gas thermometer).

13.4   THE UNIVERSAL GAS LAW

The equations (*1*), (*2*) and (*3*) can be combined in a more general equation which we may write:

$$\frac{pV}{T} = \text{a constant} \tag{4}$$

or

$$\frac{p_1 V_1}{T_1} = \frac{p_2 V_2}{T_2} \ etc. \tag{5}$$

If we always consider one mole (one gram molecular weight) of a gas, the constant is the same for all gases and we may write equation (*4*) as follows:

$$\frac{pV}{T} = R, \quad \text{where } R \text{ is the universal gas constant.}$$

Suppose a fixed mass of gas, occupying 1 litre at 27°C, is heated to 227°C and at the same time the pressure on the gas is doubled. We may find its final volume by using equation (*5*), but first we must convert the temperatures to the Kelvin scale.

$$27°C = 300 \text{ K} \quad \text{and} \quad 227°C = 500 \text{ K} \quad \text{and} \quad p_2 = 2p_1;$$

$$\frac{p_1 V_1}{T_1} = \frac{p_2 V_2}{T_2},$$

thus

$$\frac{p_1 \times 1}{300} = \frac{2p_1 \times V_2}{500},$$

hence

$$V_2 = \frac{500}{300 \times 2} = \frac{5}{6} \text{ litre.}$$

13.5   MODELS OF A GAS

The following models give an idea of how we think molecules of a gas behave. The first model consists of several marbles placed on the base of a tray or baking tin. The tray is moved horizontally by hand in short, sharp random jerks. The marbles

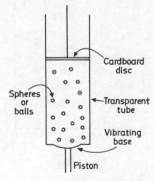

Fig. 13.7 Three-dimensional model of a gas

are seen to move, making random collisions with each other and the sides of the tray. To obtain a clear impression it is best to watch one marble carefully.

Although this model is useful, it is two-dimensional, whereas molecules move in three dimensions. Fig. 13.7 shows a better model. Some small polystyrene spheres or phosphor bronze ball bearings are contained in a vertical transparent tube. The base of the tube is connected to a piston which is driven by an electric motor. The oscillating piston causes the base to vibrate; the frequency of vibration is changed by altering the speed of the motor. An alternative arrangement is to use a loudspeaker as the base of the tube and to connect the loudspeaker to a signal generator the frequency of which can be altered.

When the base vibrates, the spheres on it are thrown into random motion in the cylinder. They collide with each other and with the walls of the container on which they thus exert a pressure. They also exert a pressure on the cardboard disc which sits on top of them. If the amplitude of vibration of the base is increased, the average energy of the spheres is increased. The spheres thus move faster and keep the disc at a higher level. This illustrates the expansion of a gas when it is heated at constant pressure. Instead of allowing the disc to rise when the spheres are made to move faster, masses can be added to the top of the disc to keep it at the same height. The masses increase the external pressure that the disc exerts on the spheres. This illustrates the increase in the pressure that a gas exerts when it is heated at constant volume.

14 Specific heat capacity

Heat is a form of energy and like any other form of energy it is measured in joules. The size of one joule is defined in terms of mechanical units.

If we take equal masses of water and oil and warm them for the same time in separate containers using similar immersion heaters, the temperature of the water will rise much less than the temperature of the oil. We say these substances have different specific heat capacities.

The specific heat capacity of a substance is the quantity of heat required to raise the temperature of one kilogram of it by 1 K. It has units of joules per kilogram per degree Kelvin (J/kg K).

It follows that if m kilograms of a substance, of specific heat capacity c, are to be raised in temperature by θ degrees K, then the heat required will be $mc\theta$ joules.

$$\text{Heat required} = mc\theta \text{ joules.}$$

It happens that for water 4200 joules of heat are required to raise the temperature of one kilogram by 1 K. The specific heat capacity of water is thus 4200 J/kg K. This value is high compared with most other substances. A great deal of energy is required to raise the temperature of water a certain amount compared with the same mass of another substance. Likewise water cools more slowly because it contains more energy than the same mass of other substances at the same temperature. Water is therefore used to fill radiators and hot water bottles.

14.1 TO MEASURE THE SPECIFIC HEAT CAPACITY OF A SOLID

The temperature of an object can be raised by supplying energy to it in various ways.

1. Mechanical energy can be supplied by allowing the object to fall through a height h, when the energy supplied is equal to mgh.
2. The energy can be obtained from another hot body of mass m, and specific heat capacity c. The energy supplied is equal to $mc\theta$, where θ is its temperature change.
3. The energy can be supplied by an electric current I, flowing for a time t, through a wire across which a potential difference of V is maintained.

$$\text{Energy supplied} = VIt.$$

The mechanical method may be used to determine the specific heat capacity of lead shot. The temperature of about 0.25 kg of lead shot is recorded. It is then placed in a wide tube about one metre long, and rubber bungs fitted at each end of the tube. The tube is smartly inverted about 100 times, the number n of inversions being counted. One bung is removed from the tube and the temperature of the lead shot again measured. The specific heat capacity c is calculated from the equation:

$$n \times mgh = mc\theta.$$

If two substances at different temperatures are mixed and come to a common temperature, the heat lost by one in cooling will be equal to the heat gained by the other, providing no heat is gained from or lost to the surroundings.

The specific heat capacity of a solid can be found by warming it to a high temperature and then quickly transferring it to a calorimeter containing cold water. The water and calorimeter receive heat from the solid and all three finally reach the same temperature. The calorimeter should either be made from a material which is a very poor conductor, in which case it can be assumed to take no heat, or from a material such as copper which is such a good conductor that it can be assumed to have the same temperature as its contents.

Details of an experiment to find the specific heat capacity of copper using a copper calorimeter are as follows. A large piece of copper, with a thread attached to it, is placed in a beaker of boiling water and left for some time. While the copper is warming to 100°C, a copper calorimeter is weighed empty and then about two-thirds full of cold water. The calorimeter is then placed inside a jacket and a thermometer placed in it.

When the piece of copper has been in the boiling water long enough to reach 100°C, the temperature of the cold water is noted, and the copper transferred from the boiling to the cold water. The mixture is gently stirred by moving the piece of copper in the water by means of the thread. The final steady maximum temperature is noted. The piece of copper is then dried and weighed.

The heat lost by the solid is then equated to the heat gained by the calorimeter and the cold water. The specific heat capacity of copper is the only unknown in the equation and can be calculated.

Example

Suppose the following readings have been obtained:
mass of calorimeter empty $m_1 = 0.1$ kg;
mass of water in calorimeter $m_2 = 0.1$ kg;
mass of piece of copper $m_3 = 0.2$ kg;
temperature of cold water $\theta_1 = 10°C$;
temperature of boiling water $\theta_2 = 100°C$;
temperature of mixture $\theta_3 = 20°C$;
specific heat capacity of water $c_1 = 4200$ J/kg K.

Heat gained by calorimeter and water = heat lost by piece of copper.
Thus $\qquad m_1 c(\theta_3 - \theta_1) + m_2 c_1(\theta_3 - \theta_1) = m_3 c(\theta_2 - \theta_3),$
hence $\qquad 0.1c \times 10 + 0.1 \times 4200 \times 10 = 0.2c \times 80$
and $\qquad\qquad\qquad\qquad\qquad 15c = 4200,$
$$c = 280 \text{ J/kg K}.$$

Although the method just described is a reasonable one, it is open to two serious errors; some hot water is carried across to the calorimeter on the hot solid and the hot solid loses heat during the transfer. Both errors can be reduced with care but not eliminated.

A better method is one based on the steady heating of a block of the solid by a heater immersed in it, as shown in Fig. 14.1.

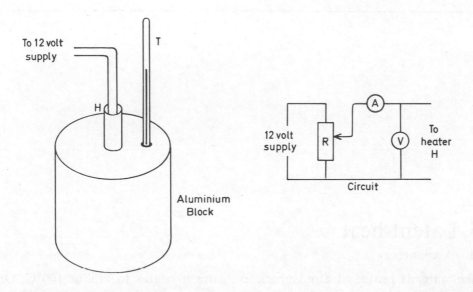

Fig. 14.1

A 12-volt immersion heater H, with a power of 24 or 36 watts, is sunk into a hole specially drilled in the solid for this purpose. It is convenient, but not essential, if the block has a mass of 1 kg. A second, smaller hole in the block contains a thermometer T. To ensure good thermal contact between the heater and block and the thermometer and block, a few drops of thin oil are placed in each hole. The block can be surrounded by an insulating jacket to reduce heat losses.

The apparatus is connected up and switched on for a known time (between 10 and 30 minutes depending on the material of the block). During this time the voltage and current are kept as constant as possible, by adjusting the variable resistor R, and their values noted. At the beginning and end of the experiment the temperature of the block is noted. The specific heat capacity c of the material of the block is worked out using the equation:

energy supplied electrically = heat gained by the block,
$$VIt = mc(\theta_2 - \theta_1),$$

where θ_1 and θ_2 are the initial and final temperatures of the block.

As well as avoiding the errors present in the method of mixtures, this method requires no calorimeter.

14.2 To measure the specific heat capacity of a liquid

If the specific heat capacity of a solid is known, the method of mixtures may be used to find the specific heat capacity of a liquid. The calorimeter is two-thirds filled with the liquid under test instead of water. The procedure and calculations are then the same as already described for finding the specific heat capacity of a solid.

However, the electrical heating method just described for a solid is again superior. The liquid (1 kg for convenience) is contained in an aluminium saucepan. The heater is immersed in the liquid, care being taken not to short the connections. The procedure is the same as that described for a solid. Strictly speaking, in working out the specific heat capacity of the liquid, account should be taken of the heat used in raising the temperature of the aluminium saucepan. However, both the mass and specific heat capacity of aluminium are likely to be much less than the values for the liquid, so no great error is involved in ignoring the heat taken in by the saucepan. Calculation of the result is then the same as for a solid.

15 Latent heat

15.1 Vaporisation

When water is heated at atmospheric pressure it begins to boil at 100°C. Once boiling begins the temperature of the water remains constant at 100°C, even though heating continues. The water is steadily absorbing energy from the heater, yet there is no increase in temperature. This energy is the energy needed to convert the water from the liquid state to the vapour state. It is used in freeing the molecules from the influence of other molecules, so that they now move independently. It is given the name **latent heat**; latent means hidden or concealed.

The specific latent heat of vaporisation of a substance is the quantity of heat required to change unit mass of the substance from the liquid to the vapour state without change in temperature.

Its value for water at 100°C is 2.26×10^6 J/kg.

15.2 Fusion

Just as latent heat is taken in when water changes to steam at the same temperature, so the same thing happens when ice melts to form water.

The specific latent heat of fusion of a substance is the quantity of heat required to convert unit mass of the substance from the solid to the liquid state without change in temperature.

Its value for ice at 0°C is 3.34×10^5 J/kg.

It follows from both definitions that if m kilograms of a substance, of specific latent heat L, change from one state to another without change in temperature, then the heat absorbed or given out will be mL joules.

Thus   energy change $= mL$ joules.

15.3   TO MEASURE THE SPECIFIC LATENT HEAT OF FUSION OF ICE

A 12-volt immersion heater should be placed in a filter funnel and surrounded by closely packed dry ice (Fig. 15.1).

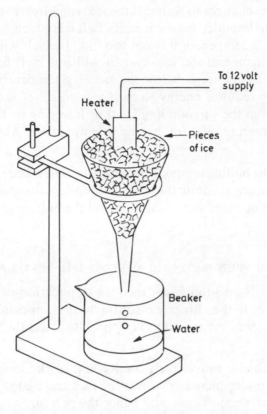

Fig. 15.1 Specific latent heat of fusion of ice

A voltmeter and ammeter are connected in the supply circuit so that the energy supplied can be determined. The immersion heater is switched on for a known time (say 3 minutes) and the mass of water produced is found by weighing the beaker before and after the experiment. A second 'control' apparatus is set up alongside and water collected from this during the same time. This tells us the mass of ice melted in the main experiment by absorption of heat from the surroundings. The difference between the two masses collected gives the true mass melted by the heater. L is calculated using the equation:

$$mL = VIt.$$

15.4   TO MEASURE THE SPECIFIC LATENT HEAT OF VAPORISATION OF WATER

A tall beaker of water is placed on a top pan balance. A 240-volt mains-operated immersion heater is carefully suspended in the water so that it does not touch the bottom of the beaker, yet is well covered with water. The heater is switched on and when the water is boiling briskly a note is made of the time and the reading of the balance. The heater is left running until a few hundred grams of water have vaporised; the time this takes will depend on the power of the heater. When sufficient water has vaporised the time and balance reading are again noted. As the immersion heater is mains operated, the voltage will be steady and dependable and the power provided by the heater will be that stated on it. The specific latent heat L is calculated from the equation:

$$mL = VIt$$

where m is the mass of water vaporised.

15.5   THE EFFECT OF IMPURITIES ON THE MELTING AND BOILING TEMPERATURES OF WATER

If a small quantity of salt is added to ice at 0°C the ice melts. Salt has the effect of lowering the freezing temperature of water, and the ice now being above the new melting temperature, changes to water. If the ice is at a lower temperature than 0°C more salt will have to be added before it melts. Salt is frequently put on icy roads; as long as the prevailing temperature is not too low, the salt will melt the ice. If the ice is at too low a temperature, however, it will not melt however much salt is added; this is because the ice is below the freezing temperature of saturated salt solution (brine). Ice requires energy to melt as a result of the addition of salt. It takes this energy from the surrounding ice thus lowering its temperature. Ice and salt are used as a freezing mixture, having a temperature which can be as low as −17°C.

Salt also raises the boiling temperature of water. The more salt added, the higher the boiling temperature is, until the water becomes saturated salt solution. The water will then hold no more salt in solution and the boiling temperature cannot be raised further.

15.6   THE EFFECT OF PRESSURE ON THE MELTING TEMPERATURE OF ICE

If a substance expands on solidifying, then the application of pressure lowers the melting temperature. If the substance contracts the opposite is the case. Water expands on freezing; pressure may thus be applied to lower the melting temperature of ice.

Ice skaters are able to move freely on account of the lowering of the melting temperature of ice under increased pressure. The knife-edge runner of the skate has a very small cross-section area, and hence the pressure under the knife-edge is very large. Together with the heat produced by friction, the net effect is the melting of the ice under the skate, and the skater moves easily through a thin film of water on top of the ice. If it is too cold, the ice does not melt, and the skate moves over the ice with difficulty.

15.7   THE EFFECT OF PRESSURE ON THE BOILING TEMPERATURE OF WATER

The effect of water boiling under reduced pressure can be shown by placing a little water in a flask, and connecting the flask to a vacuum pump via a length of thick walled rubber tubing. When the pump is switched on bubbles are soon seen rising through the water, which begins to boil at room temperature. The pump lowers the external pressure above the water thus making it easier for the water molecules to separate from each other and act independently.

As one goes higher above the earth's surface, atmospheric pressure decreases. At about 3000 m above sea-level water boils at 90°C and at 4000 m it boils at 85°C.

If the pressure is increased above that of normal atmospheric pressure the boiling temperature of water increases. A domestic pressure cooker makes use of this fact. Steam is allowed to escape from the cooker at a slow rate; most of the steam produced is thus kept inside the cooker and raises the pressure above the water. The boiling temperature of the water is thus increased above 100°C. The food inside is cooked at a higher temperature than it would be in a normal saucepan, and thus cooks more quickly. For safety reasons the pressure cooker has a valve which releases the pressure if it goes above a certain value.

15.8   EVAPORATION

Evaporation takes place from the surface of a liquid at all temperatures, whereas boiling only occurs above a certain temperature and takes place throughout the liquid. However, the nearer the liquid is to its boiling temperature the faster the rate of evaporation.

Evaporation means that the faster molecules which happen to be near the surface escape from it. If the molecules which escape are free to move away from the space immediately above the liquid (or even encouraged to do so by a stream of air) evaporation will continue until the liquid has all evaporated. This is how puddles of water left on the road after rain eventually dry up. The higher the temperature of the liquid the quicker evaporation occurs.

As evaporation means that some of the more energetic molecules of the liquid are leaving it, the average energy of the molecules left behind falls; thus the liquid falls in temperature. For example, a bottle of milk may be kept cool by wrapping the bottle in a wet cloth. Water evaporates from the cloth and that left falls in temperature. In turn it extracts heat from the milk. This is more effective if the rate of evaporation can be speeded up by standing the wet bottle in a draught.

It is unwise for a human being to stand in a draught or breeze after taking violent exercise, however warm he feels. The perspiration on his body evaporates quickly under these conditions, thus cooling his body, and making it susceptible to a chill.

15.9   VAPOUR PRESSURE

If evaporation takes place in a closed vessel the space above the liquid begins to fill with vapour. The vapour molecules move in all directions and exert pressure when they bounce off the walls of the vessel. Eventually the space above the liquid will hold no more vapour; it is said to be **saturated with vapour**. The pressure exerted by the vapour is then called **the saturated vapour pressure**. For a given temperature the saturated vapour pressure of a liquid is always the same whether or not the space above it contains another gas or vapour such as air. If the temperature of the liquid increases its saturated vapour pressure becomes higher.

16 Transmission of heat

Heat is transferred from one place to another in one of three ways:
conduction, radiation or **convection.**

16.1   CONDUCTION

If a metal spoon is left in a teacup for a short length of time the handle becomes warm. Heat travels along the spoon by means of conduction. Metals contain electrons which are very loosely attached to atoms, and are easily removed from them. When a metal is heated these 'free electrons' gain kinetic energy and move independently of the atoms. They drift towards the cooler parts of the metal thus spreading the energy to those regions.

In substances where no free electrons are present the energy is conveyed from one atom to another by collision. This is a much slower process and such substances are called poor conductors of heat.

Most metals are good conductors of both heat and electricity; free electrons being responsible for both. Substances such as wool, cotton, cork and wood are bad conductors. A number of materials lie between these extremes. The best saucepans are made of copper as heat is most rapidly conducted through the metal. Many materials are poor conductors of heat because they trap tiny pockets of air between their fibres, and air, like all gases, is a poor conductor of heat. Textiles and glass fibre are examples. Glass fibre is frequently used as lagging for attics and hot water tanks.

16.2   COMPARISON OF THERMAL CONDUCTIVITIES

Similar rods, made of different materials, pass through corks inserted in holes made in the side of a metal tank (Fig. 16.1). The rods are first coated in wax by

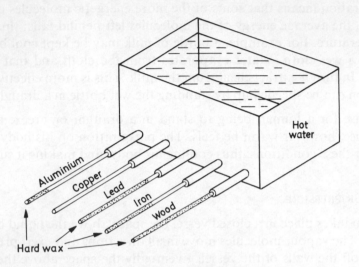

Fig. 16.1 Conduction

dipping them into molten wax and then allowing it to cool. Boiling water is then poured into the tank, so that the ends of the rods are all heated to the same temperature. After some time it is seen that the wax has melted to different distances along the rods, showing differences in their **thermal conductivities**.

In the kitchen, saucepans are made of metals such as copper or aluminium, which are good heat conductors. However, the handles of saucepans are made of insulators, such as plastic or wood, so that the utensils can be handled when hot. The handles of kettles and oven doors must be made of similar materials.

16.3   RADIATION

Both conduction and convection are ways of conveying heat from one place to another which require the presence of a material. Radiation does not require a material medium; it is the means by which heat travels from the sun through the empty space beyond the earth's atmosphere. Radiation consists of electromagnetic waves which pass through a vacuum. On striking a body these waves are partly reflected and partly absorbed, and can cause a rise in temperature.

The rate at which a body radiates depends on its temperature and the nature of its surface. For a given temperature, a body radiates most energy when its surface is dull black and least when its surface is highly polished. A comparison of the radiating powers of different surfaces may be made using a Leslie cube. This is a hollow metal cube, each side of which has a different surface; one is dull black, one highly polished, another may be shiny black and the fourth painted white. The cube is filled with hot water and a thermometer with a blackened bulb placed at the same distance from each face in turn (Fig. 16.2). In each case the thermometer reading is noted.

The results show that the dull black surface produces the highest reading and

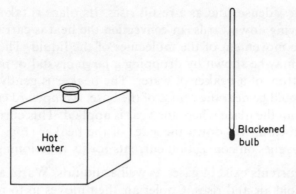

Fig. 16.2 Radiation – Leslie's cube

the highly polished one the lowest. This indicates that the dull black surface is the best radiator and the highly polished one the worst.

The absorbing powers of different surfaces may be compared using a small electric fire (Fig. 16.3).

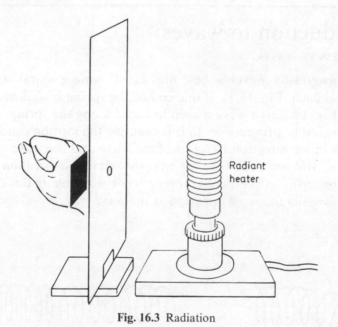

Fig. 16.3 Radiation

The fire is placed about 10 to 15 cm behind a heat insulating screen, faced with a polished metal surface towards the fire. The screen has a hole in it 2 or 3 cm in diameter. If a piece of aluminium foil is attached to the back of one's hand by damping it, it is found that the hand may be held over the hole in the screen without discomfort. If the foil is now painted over with lamp black (or matt black paint) and the hand again placed over the hole, the hand has to be removed after a few seconds or it becomes burnt. This shows that a dull black surface is a good absorber of heat as well as a good radiator, whereas a highly polished surface is poor in both respects.

Radiation from the sun is mostly in the form of visible light and infra-red rays. These pass through glass and hence may reach the ground and plants inside a greenhouse, which absorb them. These objects also radiate, but due to their relatively low temperature, the infra-red rays they emit are of longer wavelength and cannot penetrate the glass. The energy is thus trapped inside the greenhouse.

16.4 CONVECTION

When a vessel containing a liquid is heated at the bottom, the liquid in that region

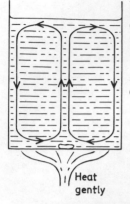

becomes warm, less dense, and as a result rises. Its place is taken by cooler, more dense, liquid moving downwards. **In convection** the heat is carried from one place to another by the movement of the molecules of the liquid. The existence of convection currents may be shown by dropping a large crystal of potassium permanganate to the bottom of a beaker of water. The beaker is gently heated under the crystal, which should be near the centre of the base. An upward current of coloured water will rise from the place where the heat is applied. This current spreads out at the surface and then moves down the sides of the beaker (Fig. 16.4). A domestic hot water system relies on convection currents for its functioning.

Fig. 16.4 Convection

Convection currents exist in gases as well as liquids. Warm air, for example, is less dense than cold air and rises. Cooler air then moves in to replace it. It is this process which is responsible for sea breezes towards the end of a warm summer's day. The relatively dark land absorbs more heat than the sea. Its specific heat capacity is less than that of water. As a result of both of these facts it reaches a higher temperature than the sea towards the end of a summer's day. The land warms the air over it which rises and is replaced by the cooler air from over the sea. Late on a clear summer's night the reverse is likely.

17 Introduction to waves

17.1 PROGRESSIVE WAVES

The idea of **progressive waves** is best illustrated using a spiral spring or slinky stretched on a bench (Fig. 17.1). If one end of the spring is shaken at right angles to its length (Fig. 17.1(*a*)) a wave is seen to travel along the spring. As the wave is moving it is said to be progressive. In this example the motion causing the wave is at right angles to the direction of travel of the wave and the wave is termed **transverse**. In Fig. 17.1(*b*) one end of the spring is shaken in the direction of the spring's length and a concertina effect travels the length of the spring. In this case the motion causing the wave is in the same direction as the wave travels and the wave is called **longitudinal**.

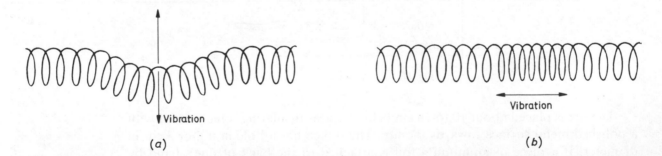

(a)   *(b)*

Fig. 17.1 Waves along a spring: (*a*) transverse; (*b*) longitudinal

Examples of transverse waves are television, radio, heat, light, ultra-violet, X-rays, γ-rays (all electromagnetic waves) and water waves. In electromagnetic waves the electric and magnetic disturbances are at right angles to the direction the wave travels. At any moment a transverse wave has the shape of a sine wave and can be drawn as such (Fig. 17.2).

Sound, on the other hand, is a longitudinal wave motion. The molecules of the material through which the sound travels, vibrate to and fro along the direction of motion of the wave. At a given time each molecule is at a different point in its motion. A longitudinal wave may be represented by a sine curve, but it must be remembered that although the y direction still represents the size of the displacement, the direction of the displacement is in fact parallel to the direction of travel of the wave.

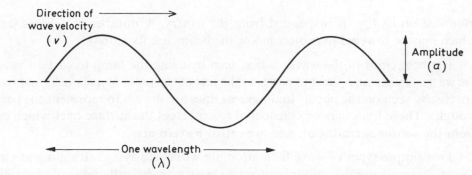

Fig. 17.2 Sine curve representing wave motion

In Fig. 17.2 the **amplitude of the wave** is denoted by *a*. The amplitude represents the maximum displacement of the wave from the zero position. The distance between corresponding points on two successive waves is known as the **wavelength λ**. The number of waves produced every second by the source is called the **frequency f**. The length of wave motion produced every second by the source is the number of complete waves produced (f) multiplied by the length of each wave (λ). This product is clearly the speed v of the wave motion.

Thus $$v = f\lambda.$$

It is usual to measure the speed in metres per second, the frequency in hertz and the wavelength in metres.

17.2 THE RIPPLE TANK

A good understanding of the properties of all types of waves can be obtained from studying the behaviour of water waves in a ripple tank. The construction of a ripple tank is shown in Fig. 17.3.

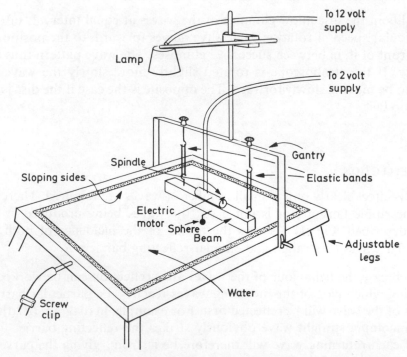

Fig. 17.3 A ripple tank

The apparatus consists of a shallow tray with sloping sides to reduce reflection. The tray is mounted on four legs, each of which can be adjusted in length by a screw foot. In this way the tray can be levelled. A gantry stands above one end of the tank. On this gantry is a post on which a lamp is fixed. A beam, with a motor mounted on its top, is suspended from the gantry. A number of spheres, each of which may be lowered to project below the beam, are fixed to it.

The behaviour of the waves is best seen by using the lamp to cast a shadow of the water surface on a sheet of paper at the feet of the tank. Wave peaks and troughs are clearly seen on the paper. In diagrams lines are drawn to represent the peaks or troughs. These lines may be considered to represent the surface over which energy from the source spreads out, and are called **wavefronts**.

Two simple types of wave formation are worth studying; straight and circular waves. A small number of straight waves may conveniently be made by rolling a short length of dowel rod to and fro on the bottom of the tank. A few circular waves may be produced by touching the water surface with a pencil the required number of times.

For many experiments it is necessary to produce a continuous series of waves of one or other type. This is done by connecting the motor to a suitable power unit (often 2v d.c.) whereupon the spindle which is unevenly weighted, sets the beam bouncing up and down. The speed of the motor determines the frequency of the waves. If straight waves are required the lower edge of the beam should be a few millimetres below the water surface; if circular waves are wanted then the beam should be raised clear of the surface but one or more spheres turned down so that they are partly submerged by the same amount.

When the motor is producing a continuous series of waves it is often very difficult, if not impossible, for the eye to follow them across the paper below the tank. If this is so, it is helpful to use a hand stroboscope to 'freeze' the picture of the waves. A typical stroboscope is shown in Fig. 17.4. The handle is held in one hand while the disc is rotated using one finger of the other hand. The rotating disc is placed between the face and the wave pattern on the paper. The speed of rotation of the disc is increased until the waves appear stationary.

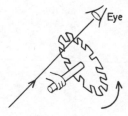

The slits in the disc allow glimpses of the waves at equal intervals of time. At one particular speed of rotation each wave moves forward, to the position of the wave in front of it, in between successive glimpses. The wave pattern thus appears stationary. If the stroboscope is rotated slightly more slowly the wave pattern appears to be moving slowly forward. The opposite is the case if the disc is rotated slightly too fast.

Fig. 17.4 Hand stroboscope

17.3   REFLECTION

To see this effect clearly only a small number of waves are required. Therefore it is best if the ripple tank motor is not used; the waves being produced in the way already described. Fig. 17.5 shows the results of the incidence of both straight and circular waves on straight and circular reflecting barriers R.

In each case the behaviour of the waves after reflection can be worked out by considering which part of the incoming waves reaches the particular barrier first. This part of the wave will be reflected first. For example, in diagram (d), the centre of each incoming straight wave obviously strikes the reflecting barrier first. The centre of each returning wave will therefore be in front, giving the curved shape shown.

In diagrams (c) and (e) the waves after reflection are approximately circular and converge to a point. This point is the image formed by waves from the point source (a distant source in (c)) after reflection at the barrier. In (d) the reflected waves appear to come from a point behind the barrier. This point represents a virtual image of the source.

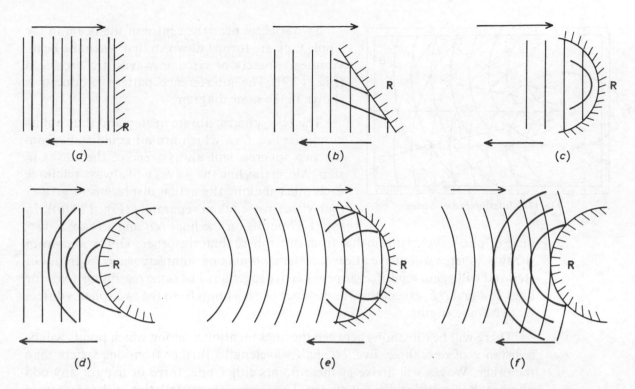

Fig. 17.5 Reflection of waves

17.4 REFRACTION

If water waves come to a region where their speed changes abruptly, then their direction of travel may change abruptly. The best way of achieving this change in speed is to alter the depth of water by placing a sheet of glass in the tray, since it is found that if the water is shallower the speed is reduced. This can be seen as waves approach a sloping beach. The body of the waves slows, but the crest is less affected and falls forward, causing the waves to break.

The sheet of glass should be placed in the tray so that it is covered by water to a depth of only one or two millimetres. A little experimentation will give the best depth. Straight waves, produced by either the beam or dowel rod, reach the leading edge of the glass plate at an angle (Fig. 17.6). As soon as part of the wave comes

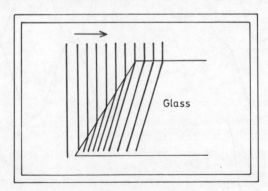

Fig. 17.6 Refraction of waves at a plane boundary

over the shallow region it slows down. It is clear that the end of the wave which has been in the shallow region longer will fall behind the other end. Thus the direction of the wavefront changes as shown.

17.5 INTERFERENCE

If water waves from two sources with the same frequency and similar amplitude meet, complete calm can be seen on some regions of the water surface. The waves are said to **interfere**.

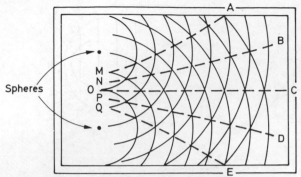

Fig. 17.7 Interference of waves

Two spheres near the centre of the beam in the ripple tank are turned down so that when the beam vibrates two sets of circular waves are produced (Fig. 17.7). The interference pattern produced is shown in the same diagram.

The two spheres vibrate in step and thus points along the line *OC*, which are all equidistant from the two spheres, will always receive the waves in step. Along this line the waves will always **reinforce** each other making the actual displacement greater than either wave taken separately (Fig. 17.8(*a*)). In Fig. 17.7 points on the lines *NB* and *PD* are a distance of one wavelength further from one sphere than the other. Thus waves from the two spheres will arrive at each of these points one complete wavelength out of step and will again reinforce, as peaks arrive together. The same reasoning holds for lines *MA* and *QE*, except that the distance of each point from the two spheres differs by two wavelengths.

There will be directions between the lines mentioned along which points will be a distance of one, three, five, *etc.* half wavelengths further from one sphere than the other. Waves will arrive at these points either one, three or five (or any odd number) half wavelengths out of step. Thus complete **cancellation** of the two sets of waves will always occur at these points (Fig. 17.8(*b*)).

If the wavelength is reduced a smaller path difference is required for the waves from the two sources to be out of step by one, two, three, *etc.* complete wavelengths. Thus the lines in Fig. 17.7 come closer together and more regions of addition and cancellation of waves are seen in the tray. The reverse is true if the wavelength is increased.

If the two sources of waves are brought closer together the lines in Fig. 17.7

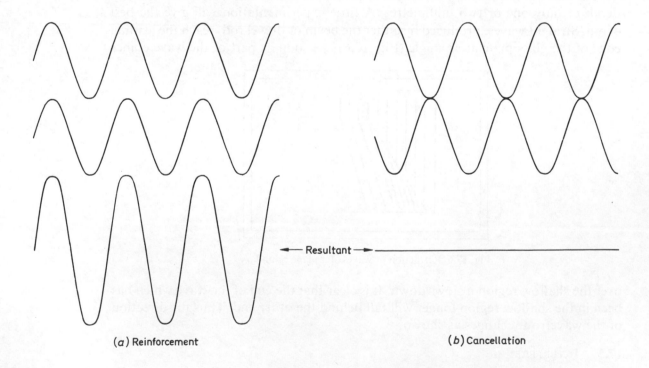

(*a*) Reinforcement

(*b*) Cancellation

Fig. 17.8 Addition of waves: (*a*) in step – reinforcement; (*b*) out of step – cancellation

become further apart and fewer regions of addition and cancellation of waves are seen in the tray. If the sources are made further apart the reverse is true.

17.6 DIFFRACTION

If straight water waves are passed through a narrow gap in a barrier or past a small object some bending of the waves round the edges of the barrier or the object is noticed. Thus some change in the direction of travel of the waves occurs round these edges. This effect is called **diffraction**.

For diffraction to be obvious the size of the gap or object has to be about the same as the wavelength of the waves. In Fig. 17.9(*a*) the gap between the barriers is

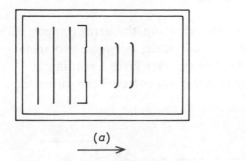
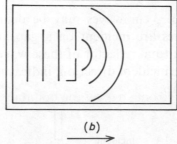
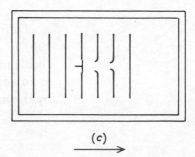

(a)   (b)   (c)

Fig. 17.9 Diffraction of waves: (*a*) wide gap; (*b*) narrow gap; (*c*) small object

much greater than the wavelength·and little bending occurs. In Fig. 17.9(*b*) the gap size and the wavelength are about the same and the bending is marked. In fact the waves become circular after passing through the gap and look the same as if the gap were replaced by a point source of waves. In Fig. 17.9(*c*) where the object is about the same width as the wavelength the bending is again very noticeable.

The interference experiment described in the last unit may be carried out by allowing a series of straight waves to pass through two gaps in a barrier. After passing through the waves spread out in a semicircular fashion; that is they are diffracted, and produce the same interference pattern as was obtained with the two vibrating spheres. It should be noted that interference only occurs in this version of the experiment because diffraction takes place at the narrow gaps. If diffraction did not take place the waves would never meet.

A clear understanding of the behaviour of water waves discussed in this unit will make it easier to understand the properties of light and sound waves discussed in later units, as they behave in a similar way.

17.7 3 CM WAVES (MICROWAVES)

These may be used to show many of the effects just discussed in connection with water waves. The apparatus consists of a transmitter T and a receiver R, which has a galvanometer built into it. If the transmitter is pointed at a wall, a sheet of hardboard or a metal plate, so that the waves strike normally,.it is found that the reflected waves return along the reverse path to the incident waves. The maximum reading is obtained on the galvanometer when the receiver is alongside the transmitter and parallel to it (Fig. 17.10(*a*)).

In diagram (*b*) the reflector is arranged so that the waves strike it at an angle to the normal. The maximum reading on the galvanometer shows that the waves are reflected through an angle such that $i = r$. The results are the same as those obtained in Fig. 17.5(*a*) and (*b*) respectively with water waves.

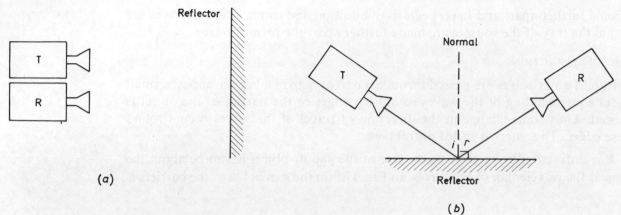

Fig. 17.10 Reflection of 3 cm waves

Diffraction of 3 cm waves may be demonstrated using the arrangement in Fig. 17.11. Waves are incident on a gap about 1 cm wide, between two metal sheets (metal is a total reflector of these waves). The receiver gives a reading at *Y*, at an angle to the incident direction, indicating that the waves have been diffracted by the narrow gap.

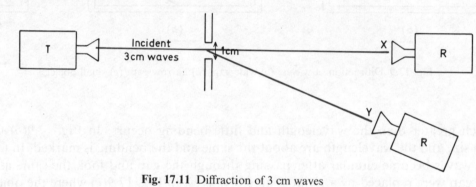

Fig. 17.11 Diffraction of 3 cm waves

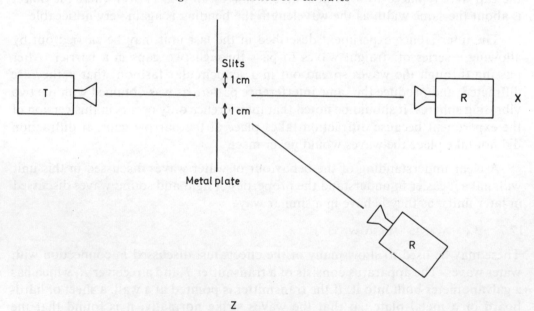

Fig. 17.12 Interference of 3 cm waves

Fig. 17.12 shows an arrangement in which 3 cm waves fall on two narrow gaps in metal plates. When the receiver is moved between *X* and *Z* it shows directions in which a relatively large wave is received. In between these directions little energy reaches the receiver. These results indicate an interference pattern similar to that obtained using water waves.

18 The passage and reflection of light

In some light experiments it is more convenient to consider the behaviour of **rays** rather than wavefronts. Rays are lines drawn at right angles to wavefronts and thus represent the direction in which the wave is travelling. A ray box is a device for producing rays; one version is shown in Fig. 18.1.

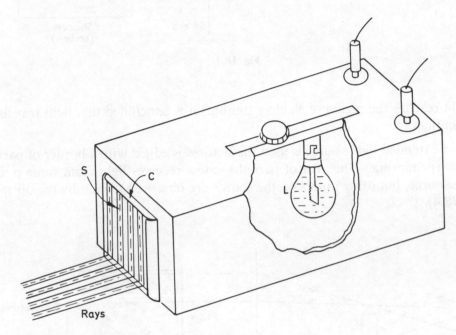

Fig. 18.1 Ray box

A small filament lamp L is enclosed in a box with a cylindrical convex lens C and a 'comb' S containing parallel slits in front of it. A diverging, converging or parallel beam of rays (Fig. 18.2) may be obtained by moving the lamp L.

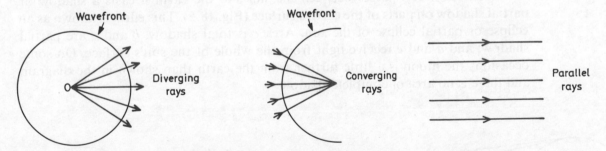

Fig. 18.2 (*a*) Diverging rays; (*b*) converging rays; (*c*) parallel rays

18.1 RECTILINEAR PROPAGATION

If a ray box is placed on a sheet of white paper and switched on, a ray is obtained through each slit of the comb. This ray is actually a very narrow beam of light with straight sharp edges. It can be said that light travels in straight lines (**rectilinear propagation**) in this case. The existence of shadows and eclipses of the sun and moon is further evidence that light travels in straight lines.

When an obstacle is placed in the path of light coming from a point source the shadow formed on the screen is uniformly dark and has sharp edges (Fig. 18.3). As

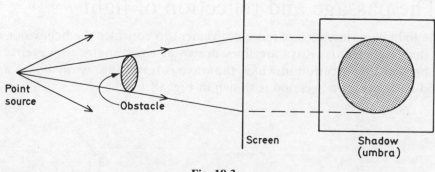

Fig. 18.3

no light reaches the region of shadow (umbra) it is concluded that light travels in straight lines.

If an extended light source is used the shadow is edged with a border of partial shadow (penumbra). The area of partial shadow receives light from some points on the source, but other points on the source are obscured from it by the obstacle (Fig. 18.4).

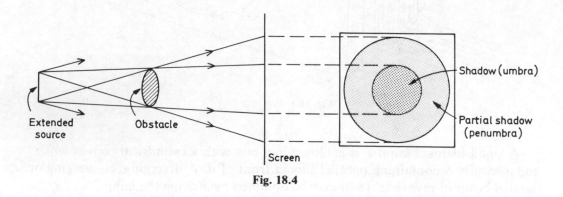

Fig. 18.4

When the moon passes between the sun and the earth it casts a shadow or partial shadow on parts of the earth's surface (Fig. 18.5). This effect is known as an eclipse or partial eclipse of the sun. Area *c* is total shadow, *b* and *d* are partial shadow, and *a* and *e* receive light from the whole of the sun's surface. On some occasions the moon is a little further from the earth than shown in the diagram and there is no area of complete shadow.

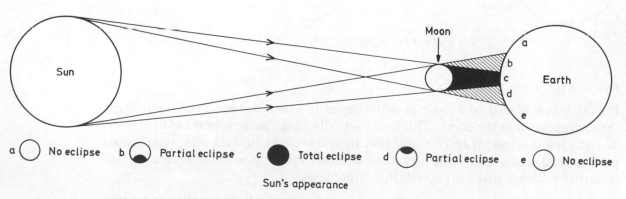

Fig. 18.5 Eclipse of the sun

The pin-hole camera relies on the fact that light travels in straight lines to produce a clear image. It is a very simple version of a camera, invented well before lenses were used to produce images (Fig. 18.6(*a*)).

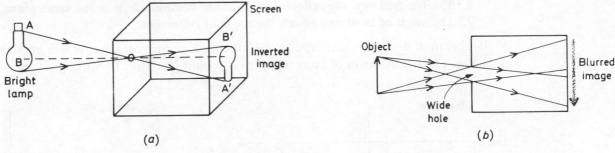

(*a*)   (*b*)

Fig. 18.6 Pinhole camera

A pin-hole camera may be made by removing the back of a small cardboard or metal box, and replacing it with a piece of semi-transparent paper (the screen). A pin-hole is punched in the side of the box opposite the screen. When the hole is held towards a bright lamp, such as a carbon filament lamp, in a darkened room, an inverted image of the lamp filament can be seen on the screen. A narrow beam of light from point *A* on the source enters the camera through the pin-hole and strikes the screen at *A'*. Likewise light from *B* arrives at *B'*. Narrow beams of light from all the different points on the object will fall on the screen between *A'* and *B'*. Each point on the object is thus responsible for a point of light on the screen and a complete inverted image is seen. This image is formed on a screen and is said to be **real**. An image which cannot be shown on a screen, but only seen by the eye is called **virtual**.

If the pin-hole camera is moved closer to the lamp the image becomes bigger. The small hole means that little light enters the camera. A larger hole would improve this, but would lead to blurring of the image (Fig. 18.6(*b*)), unless a lens was used. Thus the exposure time needed to produce a picture on a film using a pin-hole camera is long.

18.2 REFLECTION AT A PLANE SURFACE

Reflection of light may be examined by using a plane mirror supported vertically on a sheet of white paper. A line *XN*, the normal, is drawn at right angles to the mirror surface so that *N* is near the centre of the mirror (Fig. 18.7). Further lines

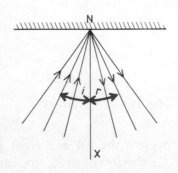

Fig. 18.7 Reflection at a plane mirror

are drawn from *N* so that they are inclined at angles such as 20°, 30°, *etc.* to *XN*. A ray box is now placed so that a single ray follows one of the drawn lines. The position of the reflected ray is marked with dots. The ray box is moved to each of the lines in turn and the procedure repeated. In each case the angle of incidence *i* of the ray is noted and the corresponding angle of reflection *r* measured. Within the limits of experimental accuracy the two are found to be equal.

The laws of reflection at plane surfaces are summarised as follows:

1. **The incident ray, the reflected ray and the normal all lie in the same plane.**
2. **The angle of incidence equals the angle of reflection.**

If a beam of divergent rays is reflected from a plane mirror, the rays will appear as in Fig. 18.8(*a*). A beam of convergent rays would give Fig. 18.8(*b*).

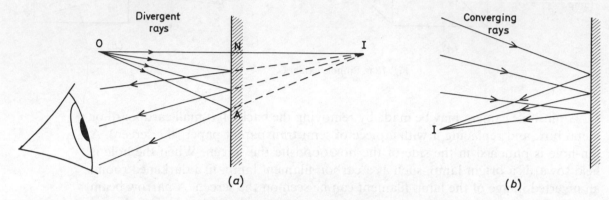

Fig. 18.8 (*a*) Diverging rays; (*b*) converging rays

In Fig. 18.8(*a*) the image *I* is the point from which rays entering the eye appear to have come. These rays have been reflected according to the laws just stated and consideration of the triangles *ONA* and *INA* will show that they are similar. Thus the distance *IN* is the same as the distance *ON*; that is the image is as far behind the surface of the mirror as the object is in front of it. As the rays do not actually come from *I* the image is virtual. In Fig. 18.8(*b*) the reflected rays cross at the point *I* to form a real image there.

If a person looks at the image of an object in a mirror he notices that it appears the wrong way round; it is said to be **laterally inverted**. Figure 18.9 shows how this

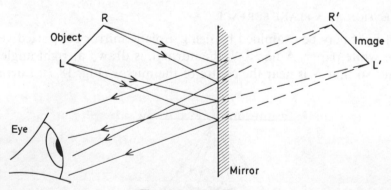

Fig. 18.9 Lateral inversion

comes about. If *L* and *R* represent the left and right sides of the object viewed directly, it will be seen that *L'*, the image of *L*, appears on the right side of the image in the mirror.

18.3 REFLECTION AT A CURVED SURFACE

Reflection of light at curved surfaces follows the same laws as reflection at plane surfaces. Fig. 18.10 shows divergent, parallel and convergent beams of rays incident on concave and convex spherical mirrors. In each of cases (*c*) and (*e*) real images are formed at the point where the rays cross after reflection at the mirror.

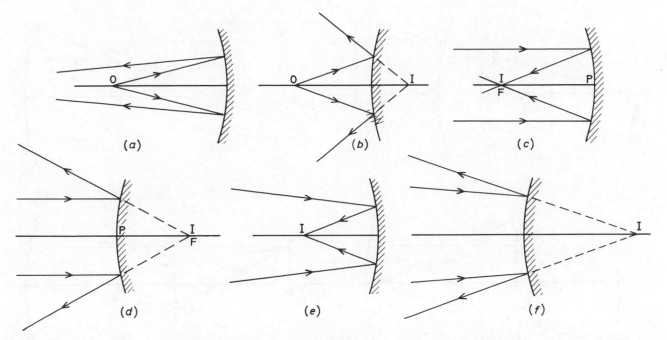

Fig. 18.10 Reflection at a curved mirror

In (*b*) and (*d*) the rays diverge after reflection, but appear to have come from a point behind the mirror in each case. A virtual image is said to exist at this point; as rays do not actually cross here it cannot be shown on a screen. Whether a real or virtual image results in (*a*) and (*f*) depends on the curvature of the mirrors and the amount of divergence or convergence of the incident beam of rays. In every case each ray is reflected according to the laws of reflection already stated, care being taken to consider the normal at the point the ray strikes the mirror.

The line drawn at right angles to the mirror surface at its centre is called the **principal axis** of the mirror. The point of intersection of the principal axis with the mirror surface is known as the **pole *P*** of the mirror.

The point *F*, in diagram (*c*), through which the rays which are parallel to the axis pass after reflection is called the **focal point** or **principal focus** of the mirror. As rays actually pass through *F*, the focal point is real. The distance *FP* is its **focal length *f*.**

The point *F*, in diagram (*d*), from which the incident rays which are parallel to the axis appear to come after reflection is called the **focal point** of the mirror. Rays do not pass through *F*, but only appear to come from *F* after reflection. The focal point is thus virtual. The distance *FP* is its **focal length *f*.**

Each of the mirrors in Fig. 18.10 are part of the surface of a sphere. The centre of the sphere is called the **centre of curvature** of the mirror. The radius of the sphere is called the **radius of curvature *r*** of the mirror.

18.4   CONSTRUCTION OF RAY DIAGRAMS

A point on an image can be located by the point of intersection of two reflected rays, which have come from the same point on the object; for example, the tip. It is generally simplest to use any two of the following four rays:

1. A ray passing through the centre of curvature *C* is reflected back along its own path.
2. A ray parallel to the principal axis is reflected through the focal point *F*.
3. A ray through the focal point is reflected parallel to the principal axis.
4. A ray incident at the pole is reflected back, making the same angle with the principal axis.

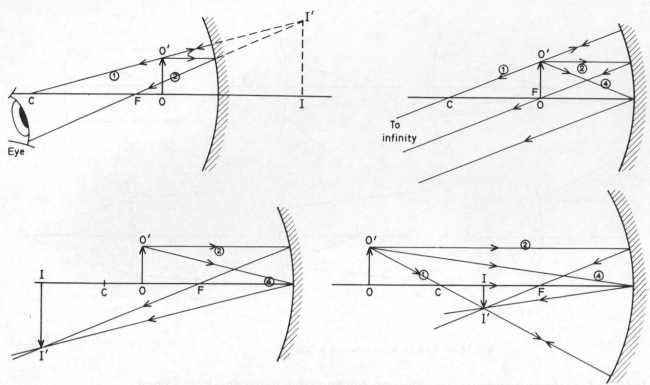

Fig. 18.11 Ray diagrams for a concave mirror

Ray diagrams are most conveniently drawn to scale on squared paper. The height chosen for the object does not matter as the height of the image will always be in proportion. Fig. 18.11 gives ray diagrams showing the image formed by a concave mirror for different positions of an object with one end on the principal axis. In all cases the magnification is given by:

$$\text{magnification} = \frac{\text{height of image}}{\text{height of object}} = \frac{II'}{OO'}$$

and magnification is equal to $\dfrac{\text{image distance}}{\text{object distance}}$.

A similar procedure is followed when constructing ray diagrams for a convex mirror (Fig. 18.12). Unlike the concave mirror, which can produce either real or

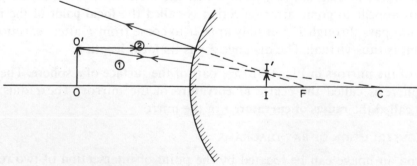

Fig. 18.12 Ray diagrams for a convex mirror

virtual images according to the position of the object, the convex mirror gives virtual images only. These are erect and smaller than the object and are formed between the mirror and its focal point. In this case it is simplest to use any two of the following four rays:

1. A ray incident in the direction of the centre of curvature is reflected back along its own path.

2. A ray parallel to the principal axis is reflected directly away from the focal point.
3. A ray incident in the direction of the focal point is reflected parallel to the principal axis.
4. A ray incident at the pole is reflected back, making the same angle with the principal axis.

For spherical mirrors – both concave and convex – it can be shown that:

$$\frac{1}{u} + \frac{1}{v} = \frac{1}{f}$$

where u = object distance from mirror,
    v = image distance from mirror,
    f = focal length.

Virtual distances carry a negative sign; this includes the focal length of a convex mirror.

19 Refraction of light

The laws of refraction are:

1. **The incident and refracted rays are on opposite sides of the normal at the point of incidence and all three are in the same plane.**
2. **The ratio of the sine of the angle of incidence to the sine of the angle of refraction is a constant.**

19.1 REFRACTION AT A PLANE BOUNDARY

The refraction of rays of light can be studied using a ray box. We will consider what occurs when a single ray of light strikes a plane air to glass boundary (Fig. 19.1).

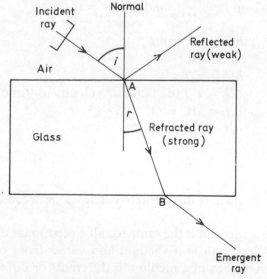

Fig. 19.1 Refraction at a plane boundary

The glass block stands on a sheet of white paper. A line at right angles to the boundary to be used is constructed near the middle of the face (a normal). It is arranged that the single ray strikes the glass block at A at various angles of incidence, i, in turn. The position of this ray and the emerging ray in each case are marked on the paper with dots. The angle of refraction r, within the block, is constructed by joining the points AB for each ray. The angle r is measured and, in each case, found to be less than the corresponding value of i; that is the ray always bends towards the normal. The ratio $\sin i / \sin r$ is calculated and found to be the same for all angles of incidence. The full significance of this result can be seen by studying Fig. 19.2.

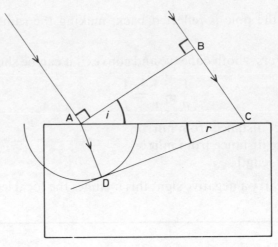

Fig. 19.2

AB represents a plane wavefront striking the boundary at an angle *i*. While the end *B* of the incoming wavefront is travelling the distance *BC* to the boundary, the light which strikes *A* will have entered the glass and will be passing through it at a lower speed. When *C* is reached the light from *A* must lie somewhere on the semi-circle shown. As the wavefront in the glass must be plane and also be passing through *C* at this moment, it must be represented by the line *CD*.

Let the velocity of light in air be v_1 and in glass v_2. Also let *t* be the time for the light to go from *B* to *C*. It will also be the time for light to travel from *A* to *D*.

Triangles *ABC* and *ADC* are both right angled triangles.

Thus
$$\sin i = \frac{BC}{AC} \quad \text{and} \quad \sin r = \frac{DC}{AC}.$$

Therefore $BC = AC \sin i$
and $AD = AC \sin r$
but $BC = v_1 t$
and $AD = v_2 t$ (distance = velocity × time);
thus $v_1 t = AC \sin i$
and $v_2 t = AC \sin r.$
Dividing gives $\dfrac{v_1}{v_2} = \dfrac{\sin i}{\sin r}.$

The ratio $\dfrac{v_1}{v_2}$ is known as the **refractive index** μ between the two materials.

The velocity in a vacuum is the same for all electromagnetic waves. The velocity in any medium is less than in a vacuum and varies from one colour to another. Hence the refractive index of a medium is different for different colours.

Thus the ratio sin *i*/sin *r* which the experiment showed to be constant is the ratio of the velocity of the light in the two materials. This is perhaps not surprising as it is the change in the velocity of light which leads to the refraction.

Fig. 19.3 shows two consequences of the refraction which occurs when light passes from water to air and thus speeds up. In diagram (*a*) the bottom of the pond appears to be at *I* and not *O*. The pond appears more shallow than it is. In diagram (*b*) the stick appears bent at the point where it enters the water.

The explanation is the same for both effects. We will first consider the case when the bottom of the pond or the end of the stick are viewed from vertically

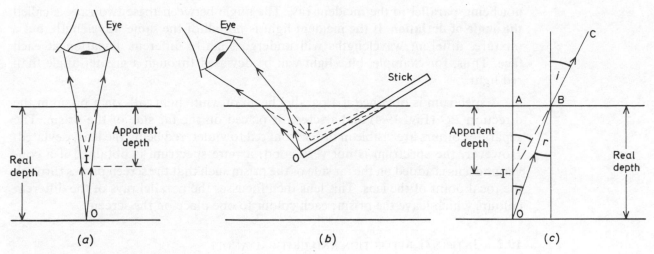

Fig. 19.3 Consequences of refraction: (*a*) apparent depth; (*b*) bent stick

above (Fig. 19.3(*c*)). The ray *OBC* in the diagrams is thus close to the normal to the water surface. The emergent ray *BC* appears to be coming from a virtual image *I*, so that *AI* is the apparent depth of the pond or end of the stick.

The refractive index μ of water is given by:

$$\mu = \frac{\sin i}{\sin r}$$

but $\qquad \angle AIB = i$ and $\angle AOB = r,$

therefore $\qquad \mu = \dfrac{\sin \angle AIB}{\sin \angle AOB}$

$$= \frac{AB}{BI} \bigg/ \frac{AB}{BO} = \frac{BO}{BI}$$

$$= \frac{AO}{AI} \quad \text{when } B \text{ is close to the normal at } A,$$

which is the case when viewing is done from directly above *O*.

Thus $\qquad\qquad \mu = \dfrac{\text{Real depth}}{\text{Apparent depth}}.$

If the viewing is done more obliquely the apparent reduction in depth is more but this formula is no longer valid.

Another example of refraction at a plane surface takes place in a prism (Fig. 19.4).

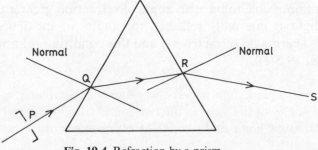

Fig. 19.4 Refraction by a prism

On entering the prism at *Q* the speed of the light is reduced and refraction takes place towards the normal. The reverse takes place at *R* as the light leaves the prism. The prism is a glass block. In the case of a rectangular glass block, the refracting faces are parallel, and the emerging ray is parallel to the incident ray. For a prism the refracting faces are not parallel and this results in the emerging ray not being parallel to the incident ray. The angle between these two rays is called **the angle of deviation**. If the incident light is not all of the same wavelength, but a mixture, different wavelengths will undergo slightly different deviation at each face. Thus, for example, blue light will be deviated through a greater angle than red light.

A spectrum is obtained if a parallel beam of white light falls on a prism in the direction *PQ* (Fig. 19.4) and a screen is placed on the far side of the prism. The separate colours are visible in order from red to violet, red being the least deviated. However, the spectrum is not very good: a pure spectrum is obtained if a converging lens is added on the far side of the prism such that the screen passes through the focal point of the lens. This lens then focusses the parallel rays of the different colours which leave the prism; each colour to one place on the screen.

19.2   INTERNAL REFLECTION AND CRITICAL ANGLE

When light passes from one medium to a more optically dense medium (*i.e.* the speed of the light is reduced) refraction occurs for all angles of incidence, together with a very small amount of reflection. But refraction does not always occur at the surface of a less optically dense medium, for example when light is passing from glass or water to air.

Consider a ray passing from glass or water to air with a small angle of incidence (Fig. 19.5(*a*)). Here we get both a refracted and a reflected ray, the latter being

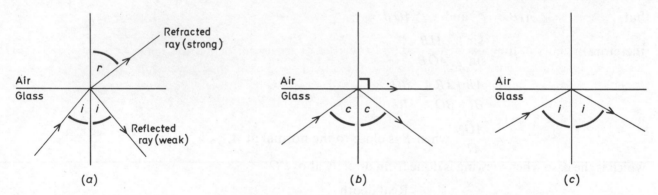

Fig. 19.5 (*a*) Refraction; (*b*) critical angle; (*c*) total internal reflection

relatively weak. If the angle of incidence is gradually increased the reflected ray becomes stronger and the refracted ray weaker, until for a certain **critical angle of incidence** *c*, the angle of refraction is just 90°. For angles of incidence greater than *c* the ray of light will not pass through. This special case is shown in Fig. 19.5(*b*).

Since it is impossible to have an angle of refraction greater than 90°, it follows that all the light is internally reflected for angles of incidence greater than the critical angle. There is no refracted ray and this condition is known as **total internal reflection** (Fig. 19.5(*c*)).

Now
$$\frac{\sin i}{\sin r} = \frac{v_2}{v_1} = \frac{1}{\mu}.$$

For critical angle $i = c$ and $r = 90°$. Hence $\sin r = 1$.

Thus
$$\frac{v_2}{v_1} = \sin c.$$

v_2/v_1 in this case is the ratio of the velocity of light in glass or water to its velocity in air (greater).

The value of the critical angle for a glass to air boundary is about 42° and for water to air about 48°.

19.3   REFRACTION AT A SPHERICAL BOUNDARY

Refraction of light at a spherical surface follows the same rules as refraction at a plane surface; that is:

1. The incident and refracted rays are on opposite sides of the normal at the point of incidence and all these are in the same plane.
2. The ratio of the sine of the angle of incidence to the sine of the angle of refraction is a constant.

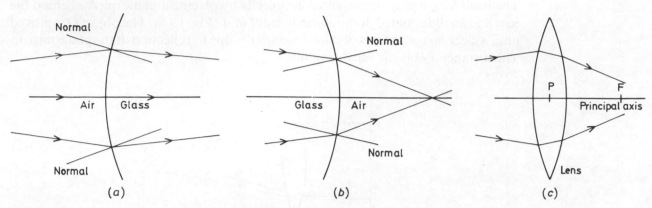

Fig. 19.6 Refraction at lens surfaces

Consider Fig. 19.6(*a*) showing three rays passing from air to glass. The three rays are all refracted according to the rules just stated; however, in this case, refraction results in the rays being brought together or focussed. If the rays leave the glass, through a second boundary curved the other way, before they have come together, further focussing occurs (Fig. 19.6(*b*)), as they bend away from the normal. The combined effect of the two surfaces is to provide us with a converging lens as shown in Fig. 19.6(*c*).

The point **F** is the point through which rays incident parallel to the principal axis pass after refraction by the lens. It is called the **focal point** or **principal focus** of the lens. The distance **PF** is known as the **focal length** *f*.

The power of the lens is defined by

$$\text{\textbf{power}} = \frac{1}{f}$$

and if *f* is measured in metres the unit of power is called a dioptre.

The behaviour of a diverging lens is shown in Fig. 19.7.

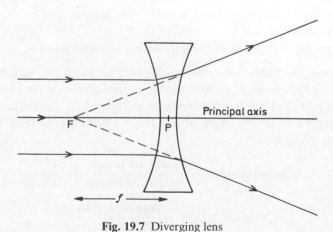

Fig. 19.7 Diverging lens

The point **F** is the point from which rays incident parallel to the principal axis appear to come after refraction by the lens. It is called the **focal point** of the lens. The distance **PF** is known as the **focal length** *f*. Because the rays do not actually pass through the focal point *F* it is virtual and the focal length *f* is negative therefore. The power is calculated in the same way as for a converging lens, but because *f* is negative the power is also negative.

19.4   To determine the focal length of a converging (convex) lens

The focal length may be measured accurately by placing a plane mirror behind the lens and an illuminated point object in front of it (Fig. 19.8). The object *O* is moved until a clear image *I* of it is obtained alongside, due to reflection at the plane mirror. The distance *PO* is the focal length *f*.

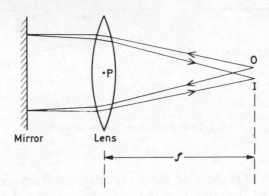

Fig. 19.8 Plane mirror method for determining the focal length of a convex lens

As the image occurs almost at the same point as the object, the light must return from the mirror almost along its incident path. This means that the light strikes the mirror normally; the light incident on the lens from the mirror is thus parallel with the principal axis.

An approximate value for the focal length of a converging lens may be found by casting the image of a distant object, such as the laboratory window in strong daylight, on a screen and measuring the distance between the lens and the screen.

19.5   Construction of ray diagrams

An image of a point on an object is formed where two rays from that point intersect after refraction. When constructing diagrams for converging lenses it is usually simplest to use any two of the following three rays:

1. A ray parallel to the principal axis which passes through the focal point after refraction.
2. A ray through the focal point which emerges parallel to the principal axis after refraction.
3. A ray through the centre of the lens which is undeviated.

Ray diagrams are most conveniently drawn to scale on squared paper. The height chosen for the object does not matter as the height of the image will always be in proportion. Fig. 19.9 gives a series of diagrams to show the type of image formed for different distances of an object from a converging lens. In all cases the magnification is given by:

$$\text{magnification} = \frac{\text{height of image}}{\text{height of object}} = \frac{II'}{OO'},$$

$$\text{magnification} = \frac{\text{image distance}}{\text{object distance}}.$$

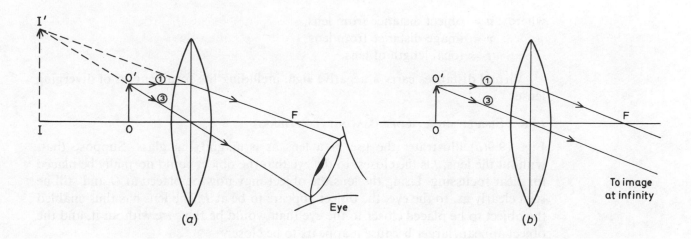

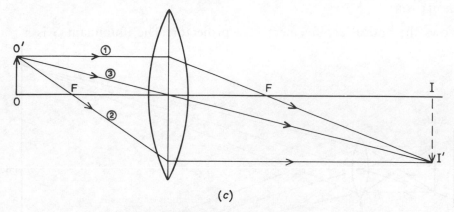

Fig. 19.9 Ray diagrams for a converging lens

When constructing ray diagrams for diverging lenses it is usually simplest to use any two of the following three rays:

1. A ray parallel to the principal axis which appears to come from the focal point *F* after refraction.
2. A ray incident in the direction of the focal point which emerges parallel to the principal axis after refraction.
3. A ray through the centre of the lens which is undeviated.

Fig. 19.10 shows the formation of an image by a diverging lens. For all

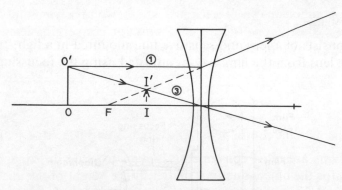

Fig. 19.10 Ray diagram for a diverging lens

positions of the object the image is virtual, erect and smaller than the object, and is situated between the object and the lens.

For both converging and diverging lenses it can be shown that:

$$\frac{1}{u} + \frac{1}{v} = \frac{1}{f}$$

where u = object distance from lens,
          v = image distance from lens,
          f = focal length of lens.

Virtual distances carry a negative sign, including the focal lengths of diverging lenses.

19.6   SIMPLE MICROSCOPE (MAGNIFYING GLASS)

Fig. 19.9(a) illustrates the use of a lens as a magnifying glass. Suppose that, without the lens, I is the closest to the eye that the object could normally be placed for clear focussing. Using the lens the object may now be placed at O and still be seen clearly as, to the eye, the object appears to be at I. The lens has thus enabled the object to be placed closer to the eye than would be the case without it, and the object appears larger because it appears to be closer.

19.7   THE PROJECTOR

Fig. 19.11 shows the optical arrangement of a projector. The illuminant S is a

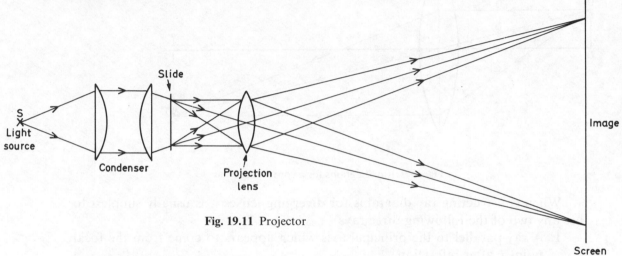

Fig. 19.11 Projector

small but very compact light source. The two condenser lenses concentrate light, that would otherwise be lost, on to the slide. The projection lens is mounted on a sliding tube so that it may be moved to and fro to focus a sharp image on the screen. The projection lens behaves as in Fig. 19.9(c). The image is inverted and thus the slide has to be inserted upside down.

19.8   THE CAMERA

The camera consists of a lens and sensitive film mounted in a light-tight box. The distance of the lens from the film may be adjusted using the focussing ring. This is

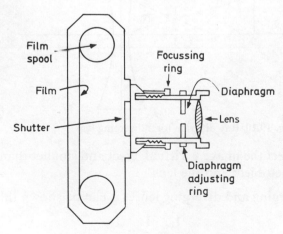

Fig. 19.12 Camera

marked in metres to show the distance at which an object will give a clear image. The amount of light entering the camera is controlled by a diaphragm of variable aperture (hole size) and the speed of the shutter. These used together allow the correct amount of light to reach the film and give the right exposure. The main features are shown in Fig. 19.12.

The ray diagram for a camera is shown in Fig. 19.9(*c*).

19.9   THE EYE

A simplified diagram of the human eye is shown in Fig. 19.13. In many ways the eye is similar to the camera. An image is formed by the lens on the sensitive retina at the back of the eye.

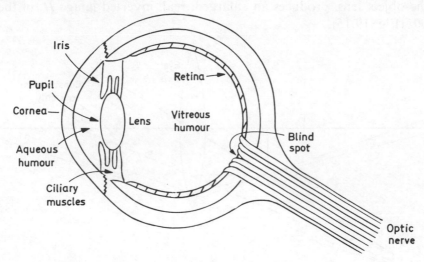

Fig. 19.13 The eye

The iris automatically adjusts the size of the circular opening in its centre, known as the pupil, according to the brightness of the light falling on it. Focussing of the image on the retina is achieved partly by refraction at the curved surface as light enters the eye, and partly by the action of the lens. The ciliary muscles vary the thickness of the lens and hence its focal length. When the muscles are relaxed the lens is thin. Unlike the camera, the position of the lens does not alter.

If the ciliary muscles weaken, the eye lens cannot be made sufficiently fat to clearly focus close objects. However, distant objects can be clearly focussed. The eye is said to suffer from **long sight**. The same defect occurs if the eyeball is too short, and it may be corrected by using spectacles containing a converging lens (Fig. 19.14(*a*)). In the case of **short sight** the muscles do not relax sufficiently and

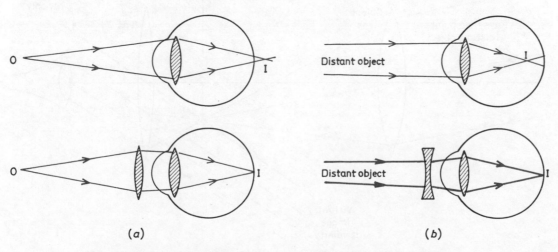

(*a*)   (*b*)

Fig. 19.14 Eye defects and their correction: (*a*) long sight; (*b*) short sight

consequently distant objects are focussed in front of the retina. Short sight also occurs if the eyeball is too long. The defect can be corrected with a diverging lens (Fig. 19.14(*b*)).

In the case of long sight bifocal spectacles are frequently worn. Each eyepiece is in two halves, the upper half often being a plane sheet of glass through which to look at distant objects and the lower half being a converging lens through which to view close objects, such as a book or newspaper. Such an arrangement removes the necessity to be continually removing and replacing spectacles.

19.10 THE COMPOUND MICROSCOPE

The compound microscope consists of two short focal length lenses. The first lens, called the object lens, produces an enlarged, real, inverted image *II'* of the small object *OO'* (Fig. 19.15).

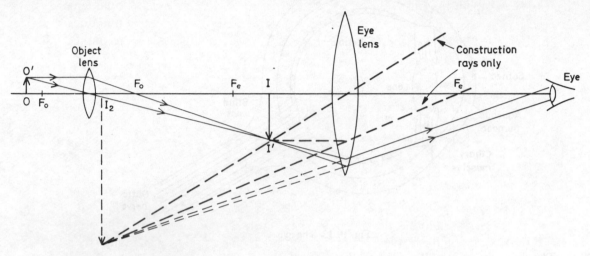

Fig. 19.15 Microscope

This image then acts as an object for the eye lens, which behaves as a magnifying glass (Fig. 19.9(*a*)) and produces an enlarged virtual image I_2, which is observed by the eye. F_o and F_e indicate the focal points of the object and eye lenses respectively. As the eye lens does not invert, the final image is still inverted.

19.11 THE TELESCOPE

A telescope is designed to increase the angle which a distant object appears to subtend at the eye; thus making the object appear larger and hence nearer.

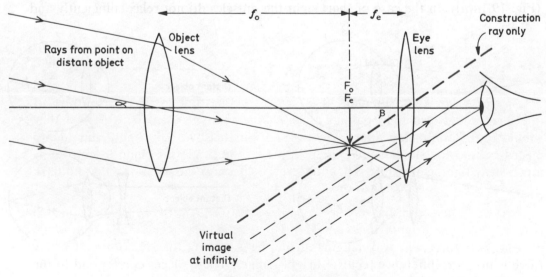

Fig. 19.16 Telescope

Fig. 19.16 shows the working of an astronomical telescope. For a distant object incoming rays are effectively parallel with each other. The object lens thus forms a real image in the plane of its focal point (F_o). The focal point (F_e) of the eye lens (magnifying glass) is usually arranged to lie in the same plane. The real image I acts as an object for the eye lens which forms a final, virtual, magnified image of infinity.

The angle subtended at the eye by the final image at infinity is very much greater than that subtended at the eye by the distant object. The ratio of these two angles is known as the **magnifying power**.

$$\textbf{Magnifying power} = \frac{\textbf{angle subtended at eye by image}}{\textbf{angle subtended at eye by object}} = \frac{B}{\alpha}.$$

For astronomical purposes it does not matter that the final image is inverted, but for use on earth the difficulty can be overcome by inserting an additional lens between the object and eye lenses. This extra lens acts purely as an erector and does not contribute to the magnifying power. The inclusion of this additional lens makes the telescope very long.

---

20 The wave behaviour of light

20.1 DIFFRACTION

The **diffraction** or spreading out which occurs when waves pass through a narrow gap or round a narrow object has already been mentioned in the unit on water waves. It was then stated that the size of the gap or object had to be about the same as the wavelength of the waves, if significant bending was to occur. Light has a wavelength (about $0.5/10^6$ m) very much less than the wavelength of water waves. The dimensions of the gap or object must be very small therefore if diffraction of light is to be observed. In practice a slit which is sufficiently narrow for the purpose can be made by drawing a sharp razor blade across the surface of a blackened microscope slide.

If the slide is held close to the eye and a light filament viewed through the slit, the filament looks much broader than it does with the naked eye, thus showing that light has been diffracted round the edges of the slit. The effect is most obvious if the filament and slit are parallel to each other. A similar effect can be obtained by partially closing one's eyes, until there is a narrow slit between the eyelids, and then looking at a light.

20.2 INTERFERENCE

Fig. 20.1 shows a simple arrangement (Young's slits) to demonstrate the **interference** of light.

S_1 and S_2 are two slits marked on a microscope slide as suggested above. If interference between light passing through the two slits is to be observed they should be 0.5 mm or less apart. They are illuminated by bright light from a lamp placed behind the single slit S; the purpose of the single slit being to cut out as much stray light as possible. The interference effect, in the form of light and dark fringes (Fig. 20.1), is observed on a screen placed a metre or so from the slits.

At certain places on the screen the light waves from the two slits will be in step (Fig. 17.8(*a*)), and bright fringes are seen. This occurs for places on the screen which are one, two, three, *etc.* wavelengths further from one slit than the other, or where there is no path difference (central bright fringe). These places correspond to the directions in Fig. 17.7 where the water waves reinforce.

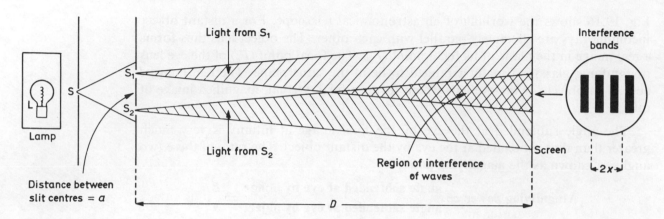

Fig. 20.1 Interference of light waves – Young's experiment

Waves from the slits arriving at places on the screen about midway between the bright fringes will be completely out of step. These places are one, three, five, *etc.* half wavelengths further from one slit than the other, and correspond to the directions of calm water in Fig. 17.7.

If the slits in Fig. 20.1 are drawn closer together the fringes on the screen are more spread out and *vice-versa*. If the lamp used produces white light, which is a mixture of wavelengths, fringes for the different colours will occur at slightly different positions on the screen. The centre fringe will be white but colours will be seen after two or three fringes to either side of this.

The equation $x = \dfrac{\lambda D}{a}$ connects the distances shown in Fig. 20.1, where x is the fringe separation, D the distance between the plane of the slits and the screen, a the separation of the slits and λ the wavelength of the light used.

20.3   THE DIFFRACTION GRATING

A **diffraction grating** consists of a large number of parallel lines ruled very close together (between 500 and 3000 lines per centimetre) on a transparent plate; it is usually produced photographically. Light passing through each of the narrow gaps between the rulings is diffracted and overlaps in the region beyond the grating. The grating thus acts like a multi-light source. An arrangement using a diffraction grating is shown in Fig. 20.2.

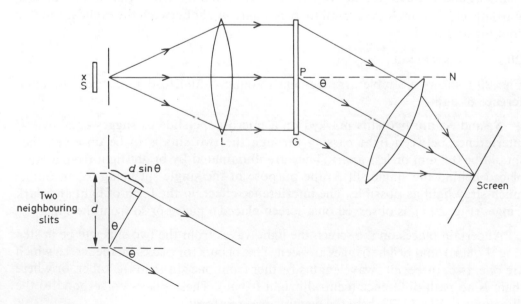

Fig. 20.2 Diffraction grating

A bright lamp S, with a colour filter in front, is placed behind a narrow slit. A converging lens L is positioned with the slit at its focus, so that a parallel beam of light is incident normally on the grating G. The emerging parallel beam in a given direction is then collected by moving a lens round from the direction PN until an image of the slit is first received on the screen. The image occurs at an angle θ such that the path difference between light rays collected from neighbouring slits is one wavelength. A second image is received when the path difference is two wavelengths *etc.*

The path difference between rays from neighbouring slits is given by $d \sin \theta$, where d is the distance between the centres of such slits. Thus when $d \sin \theta = n\lambda$ where $n = 1, 2, 3$, *etc.*, reinforcement occurs and an image is received. The values of d and θ can be measured and a value obtained for the wavelength λ.

The filter is used to produce light of approximately one particular wavelength. If white light is used images occur at different angles for different wavelengths and a series of spectra are received. The central image is white.

The apparatus may be simplified by dispensing with the two lenses. If the lamp is placed a long way in front of the grating the light incident on the grating will be parallel without using a lens to make it so. If the images are viewed directly by eye rather than on a screen the eye lens replaces the second lens.

20.4 THE ELECTROMAGNETIC SPECTRUM

The light waves discussed in recent pages are just one small part of a family of waves called the **electromagnetic spectrum**. This family includes radio and television waves, infra-red rays, light, ultra-violet light, X-rays and γ-rays. All members travel with a velocity of 3×10^8 m/s in space. Their difference lies in their frequency and wavelength. Radio waves lie at the long wavelength–low frequency–end of the spectrum and γ-rays at the short wavelength–high frequency end. Because of their differing frequencies and wavelengths, different regions of the spectrum exhibit different properties. Some of these properties are summarised in Fig. 20.3.

Wavelength-$\lambda(m)$	Type	Production	Reflection	Refraction	Diffraction and interference
$> 10^{-4}$	Radio	electrons oscillating in wires	ionosphere	atmosphere	two stations
$7 \times 10^{-7} \rightarrow 10^{-4}$	Infra-red	hot objects	metal sheet		
$4 \times 10^{-7} \rightarrow 7 \times 10^{-7}$	Visible	very hot objects	metal sheet	glass	grating
$10^{-9} \rightarrow 4 \times 10^{-7}$	Ultra-violet	arcs and gas discharges			
$10^{-12} \rightarrow 10^{-8}$	X-rays	electrons hitting metal targets			crystals
$< 10^{-10}$	γ-rays	radioactive nuclei			

Fig. 20.3

Radio waves form the basis of all long distance 'wireless' communication. The behaviour of these waves is determined largely by the presence round the earth of the ionosphere. This region of ionised gas, at heights between 80 and 400 kilometres above the earth's surface, acts as a mirror for radio waves of many frequencies. Waves of long wavelength and low frequency are reflected by this layer and such waves are thus very useful for communications round the earth's surface. Waves of short wavelength and high frequency (>30 MHz) are able to penetrate the ionosphere and are used for all communications with artificial satellites. Radio waves can pass comparatively easily through brickwork and concrete.

Any object heated nearly to red heat is a convenient source of infra-red radiation. The radiation is readily absorbed (and also emitted) by objects with rough black surfaces, but strongly reflected by light polished ones. This radiation may therefore readily be detected by painting the bulb of a thermometer black and allowing the radiation to fall on the bulb. The thermometer reading rises.

Radiation with wavelengths just shorter than that of the violet end of the visible spectrum is termed ultra-violet radiation. It is emitted by any white hot body, such as the filament of an electric light bulb. However, in this case the radiation is absorbed by the glass envelope of the bulb. A discharge tube containing mercury vapour (with quartz envelope) is a more intense source of ultra-violet radiation. A great deal of ultra-violet radiation is emitted by the sun, but the majority of this is absorbed by the earth's atmosphere and only a small fraction reaches the surface of the earth. It is this which causes browning of the skin. One of the best known properties of ultra-violet radiation is its ability to cause substances to **fluoresce**. This is the term given to the emission of visible light by substances when ultra-violet radiation is shone on them. Examples are real diamonds, uranyl salts and paraffin oil.

The X-ray region of the electromagnetic spectrum is defined more by the method of generation of the radiation than by its precise wavelength. The method used is to cause electrons to hit a metal target. X-rays are used in medicine. Their ability to penetrate matter depends on the atomic number (see unit 28.2) of the nuclei of the material through which they pass. X-rays are comparatively easily absorbed by bone which contains calcium ($Z = 20$), whereas they pass much more readily through organic material which contains hydrogen ($Z = 1$) and carbon ($Z = 6$). In addition X-rays affect a photographic plate in a similar manner to visible light. An X-ray photograph may therefore be taken of the human leg, for example, and a bone fracture detected. X-rays are dangerous in too intense quantities. They eject electrons from the region on which they fall and can cause damage to living cells. In controlled quantities they can be used to kill off diseased tissue. X-rays like other forms of electromagnetic radiation show diffraction effects. The atomic spacing in crystals is comparable with the wavelength of X-rays, and thus X-rays are appreciably diffracted by atoms in crystals. The development of the science of X-ray crystallography has led to a much more detailed understanding of the structure of materials.

γ-rays are distinguished from X-rays in that they are emitted by the nuclei of natural or artificial radioactive materials. They are generally more penetrating, more dangerous and more difficult to screen than X-rays. They are mentioned in more detail in the unit on radioactivity (unit 28.4).

21 Sound

Sound waves differ from waves of the electromagnetic spectrum in that they are mechanical waves requiring a medium through which to pass. This may be demonstrated by placing an electric bell inside a glass jar from which the air can be removed (Fig. 21.1).

When all the air has been pumped from the jar the ringing can no longer be heard, although the hammer can still be seen striking the gong. The faint vibration which is still audible comes from the passage of sound through the connecting wires.

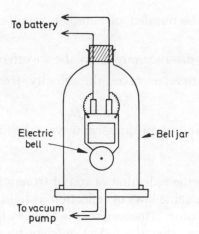

Fig. 21.1 Passage of sound

Sound waves travel through solids and liquids as well as gases. If one person places his ear near a long metal fence while a second person gives the fence a tap some distance away two sounds will be heard. The first is due to transmission through the fence; the second comes through the air. Sound travels about 15 times faster through a solid than through air. Fishermen are using transistor 'bleepers' which, when lowered into the sea, attract fish up to one mile away. The speed of sound in water is about 1400 m/s.

Sound waves are **longitudinal** rather than transverse; that is the wave particles oscillate in the same direction as the wave travels and not at right angles to it. Owing to their longitudinal nature sound waves consist of a series of **compressions** and **rarefactions**.

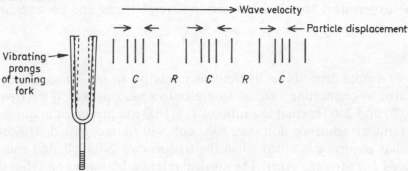

Fig. 21.2

Fig. 21.2 shows how a vibrating tuning fork sends out a sound wave. The prongs are set vibrating by striking the fork. When the right-hand prong moves to the right it pushes the layers of air in that direction to the right. These layers are thus pushed closer together; that is, they are compressed. This disturbance is transmitted from one layer of particles to the next with the result that a compression pulse or high pressure region moves away from the fork. Similarly, when the prong moves to the left, a low pressure region or rarefaction occurs to its right. Compressions and rarefactions move alternately through the air. The particle at the centre of a compression is moving through its rest position in the same direction as the wave, whilst a particle at the centre of a rarefaction is moving through its rest position in the opposite direction to the wave. At regular intervals in the material through which the sound is travelling, there will be particles undergoing exactly the same movement at the same moment. Such particles are said to be **in phase**.

As in the case of transverse waves, the distance between two successive particles in the same phase is called the **wavelength** λ.

The **amplitude** a of a wave is the maximum displacement of a particle from its rest position.

The **frequency** f is the number of complete oscillations made in one second. The unit is the hertz.

The **velocity** v is the distance moved by the wavefront in one second.

As in the case of transverse waves, the velocity, frequency and wavelength are related by the equation:

$$v = f\lambda$$

which is proved in the unit on progressive waves (see unit 17.1).

21.1   ECHOES

Echoes are produced by the reflection of sound from a hard surface such as a wall or cliff. Sound obeys the same laws of reflection as light. Thus if a person claps his hands, when standing some distance from a high wall, he will hear a reflection some time later. In order that the reflection may be heard separately from the original clap it must arrive at least 1/10 s later. As the speed of sound in air is 340 m/s, this means that the wall must be at least 17 metres away.

Echoes may be used to measure the speed of sound in air reasonably accurately. One person should make the sound and another carry out the timing. The first person claps his hands and listens for the echo from a wall. For accurate results the time interval will need to be about half a second; thus he needs to be about 100 metres from the wall. Having obtained an estimate of the time interval, he continues to clap his hands and adjusts the rate of striking until each clap coincides in time with the arrival of the echo of the previous clap. When the correct rate of striking has been achieved the second person times 30 or more intervals between claps with a stopwatch. The speed of sound is then calculated by dividing the distance to the wall and back by the interval between two successive claps. The experiment should be repeated several times and an average value calculated.

21.2   PITCH

The **pitch** of a note depends on its frequency relative to other notes. This can be demonstrated by connecting a signal generator to a loudspeaker. If the frequencies 120, 150, 180 and 240 Hz (that is a ratio of $4:5:6:8$) are produced in quick succession, the familiar sequence doh, me, soh, doh will be recognised. However, the same familiar sequence is noted when the frequencies 240, 300, 360 and 480 Hz are produced one after the other. The musical relation between notes thus depends on the ratio of their frequencies rather than their actual frequencies. An octave is always in the ratio $1:2$.

21.3   INTENSITY AND LOUDNESS

The most important factor affecting the **intensity** of a sound of given frequency is its **amplitude**. The intensity depends on the square of the amplitude and thus if the amplitude is doubled the intensity increases by a factor of four. Intensity is a measure of energy.

The **loudness** of a sound will obviously depend on its intensity. However, it also depends on the sensitivity of the human ear to sounds of different frequency.

21.4   QUALITY

A piano note can be distinguished from a trumpet note of the same pitch and loudness. The property which distinguishes them is known as **quality** or **timbre**. Musical instruments emit not only the basic or **fundamental** note but also **overtones**. An overtone (or harmonic) is a multiple of the fundamental frequency. Different instruments emit varying quantities of different overtones and it is the quantity of each present which determines the quality of the note. The note from the trumpet possesses a quality derived from the presence of strong overtones of high frequency.

21.5 INTERFERENCE OF SOUND

This can be demonstrated using two loudspeakers both of which are connected to the same signal generator. The speakers then emit waves which start in phase and have the same amplitude (coherent sources). The speakers are placed about one metre apart on a bench such that they face the same way. If one walks slowly across the room, about a metre in front of the speakers, interference can be heard. Alternatively a microphone may be connected to a cathode ray oscilloscope and used as a detector instead of the ear.

Fig. 20.1 shows how the fringes are formed. In the case of sound, S_1 and S_2 are the positions of the speakers. The waves are longitudinal, not transverse. D and a each have a value of about one metre. The line of the screen indicates the direction of movement of the ear or microphone.

21.6 STATIONARY WAVES

A **stationary wave** is formed when two progressive waves, of equal amplitude and the same frequency, meet when travelling in opposite directions. The apparatus for demonstrating this effect is shown in Fig. 21.3.

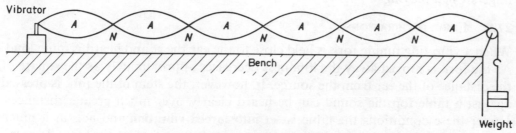

Fig. 21.3 Stationary waves

One end of a length of string passes over a pulley and is kept taut by attaching a weight to it; the other end is fixed to a vibrator. When the vibrator starts a progressive wave travels along the string to the pulley and is then reflected back. Interference occurs between the incident and reflected waves and if the frequency of the vibrator is right a standing or stationary wave is set up in the string.

The points where no vibration occurs at any time are called **nodes** N. Between each pair of nodes the string vibrates with increasing amplitude. At the centre points it reaches a maximum; these points are called **antinodes** A. Twice during a complete vibration the string is perfectly straight, and at these moments each particle in it simultaneously passes through its rest position. The distance between successive nodes or antinodes is half a wavelength. The region between two nodes is called a **segment**.

21.7 THE SONOMETER

This consists of a long sounding box with a peg at one end and a pulley at the other. One end of the wire is fixed to the peg, the other passing over a pulley to some weights. The two bridges are used to alter the effective vibrating length of the wire (Fig. 21.4).

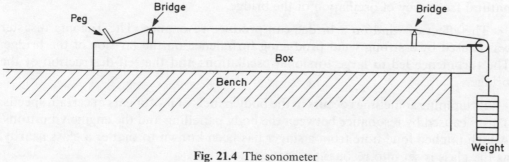

Fig. 21.4 The sonometer

When the wire is plucked in the centre waves travel to each of the two bridges and are reflected back. A stationary wave is set up and in its simplest mode the wire vibrates in one segment. As it does so it emits a note of definite frequency. The frequency *f* of this note is given by the equation:

$$f = \frac{1}{2l}\sqrt{\frac{T}{m}},$$

where *l* = the length of wire between the bridges,
       T = the tension in the wire,
       m = the mass of unit length of the wire.

The validity of this formula may be demonstrated by varying *l*, *T* or *m*, one at a time and matching the frequencies emitted by the wire with those emitted by tuning forks.

The dependence of the frequency of a vibrating string on the quantities above may be noted by watching a violinist. The instrument is 'tuned' by adjusting the tension *T* in the wires. The violin has four strings, each of a different thickness. The note emitted by a particular string is changed by altering its length with the fingers when playing.

21.8   FORCED VIBRATIONS

When a vibrating tuning fork is held close to the ear the sound heard is quite loud. As the fork is moved away the loudness decreases approximately as the square of the distance of the ear from the source. If, however, the stem of the fork is pressed against a table top the sound can be heard clearly over much greater distances. Under these conditions the table is set into **forced vibration** and acts as a much larger source. It may be considered as a large number of point sources all contributing to the loudness of the sound in the room.

Although the amplitude of vibration of the table top is much less than that of the fork, it acts over a much greater area. Energy is thus transmitted from the table top to the air at a much faster rate than it is from the tuning fork. Thus the sound lasts for a shorter time.

21.9   RESONANCE

If the body set in vibration happens to have the same natural frequency as that of whatever is causing the forced vibration, then its amplitude of vibration will rapidly increase to a very large value. This effect is called **resonance**.

For example, the best way to set a child's swing in motion is to give it a series of small pushes in time with its own period of swing. The frequency of vibration being imposed on the swing corresponds to its own natural frequency and its amplitude of swing rapidly builds up. A diver jumping repeatedly at one end of a springboard will set it into resonant vibration and so gain considerable uplift before he dives. Cases have been known where suspension bridges have been damaged by the resonant vibrations caused by marching troops. This can happen if the rate of marching happens to bear a simple numerical relationship to the natural frequency of oscillation of the bridge.

The effect of wind on a bridge can produce resonance. The Tacoma disaster was caused by a strong wind producing turbulence on the far side of the bridge. The turbulence led to large torsional oscillations and the self-destruction of the bridge.

Drumming sometimes occurs in the body panels of motor cars at certain speeds. This is caused by resonance between the body panelling and the engine vibrations. A high pitched loud note from a singer has been known to shatter a glass nearby, as the glass is set into resonant vibrations.

The action of tuning a radio set is to adjust the value of a capacitor in a circuit until the circuit has the same natural period of oscillation for electricity as that of the incoming signal. The small alternating e.m.f. set up in the aerial is then able to build up an e.m.f. of large amplitude in the tuned circuit.

Resonance may be made to occur in closed tubes using the apparatus shown in Fig. 21.5. By raising or lowering the water level it is possible to increase or decrease the effective length of the air column in the glass tube.

A vibrating tuning fork is held over the end of the tube, and the length of the column slowly adjusted until strong resonance occurs. If the length of the column is now increased resonance again occurs when the column is approximately three times as long.

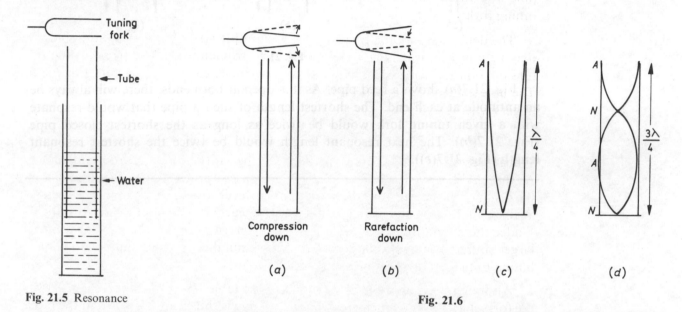

Fig. 21.5 Resonance

(a)   (b)   (c)   (d)

Fig. 21.6

Consider the first resonance position (Fig. 21.6(a)). As the prong moves downward through half a vibration it sends a compression down the tube. This compression is reflected from the water surface and returns. If it arrives back just as the prong moves up the two will be in step and reinforce as the prong is now sending a compression upwards and a rarefaction downwards. The prong has completed half a vibration while the compression has travelled a distance equal to twice the length of the column. It follows that the length of the air column is one-quarter of a wavelength (λ) as shown in Fig. 21.6(c). In the second position of resonance the length of the air column is equal to three-quarters of a wavelength (λ) as in Fig. 21.6(d).

Another way of explaining the air vibration in a tube is to regard it as resulting from the formation of a stationary longitudinal sound wave. Due to the vibration of the prong above the tube, the top of the air column must always be an antinode. The bottom of the column is a rigid boundary as far as the air is concerned and must be a node. Figs. 21.6(c) and (d) show the lowest frequency stationary waves that will fulfil these requirements. Owing to the difficulty of representing longitudinal waves diagrammatically it is usual to represent them symbolically by the use of transverse wave curves as in Figs. 21.6(c) and (d).

So far we have considered resonance in tubes closed at one end (closed pipes). Some organ pipes fall into this category (flue pipes), but others (reeds) are open at both ends (open pipes). We must now consider the setting up of stationary waves in this second category (Fig. 21.7).

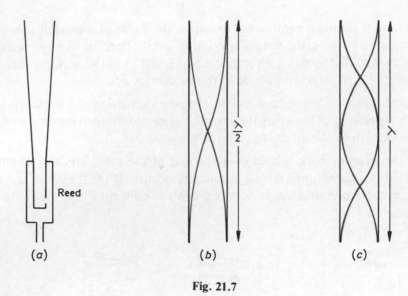

Fig. 21.7

Fig. 21.7(*a*) shows a reed pipe. As it is open at both ends, there will always be an antinode at each end. The shortest length of such a pipe that would resonate with a given tuning fork would be twice as long as the shortest closed pipe (Fig. 21.7(*b*)). The next resonant length would be twice the shortest resonant length (Fig. 21.7(*c*)).

22 Magnets

Certain materials have the property to attract iron; this property is known as **magnetism**. One such material is an iron ore called lodestone or magnetite; others are iron, steel, cobalt and nickel. Recently various alloys have been produced which can be made into very strong magnets. Alni, alcomax and ticonal are alloys of nickel and cobalt which are used for making powerful permanent magnets. Mumetal is an alloy which has been developed for the electromagnet and transformer, in which temporary magnets are used.

When a bar magnet is placed on a cork floating on water, so that it can swing in a horizontal plane, it comes to rest with its axis approximately in the north–south direction. The vertical plane in which the magnet lies is called the **magnetic meridian**. The end which points towards the north is called the north-seeking pole, or the N pole for short, and the other end the south-seeking or S pole.

If the N pole of a second magnet is brought near the N pole of the magnet on the floating cork, repulsion occurs and the cork and magnet tend to swing round. Repulsion occurs between two S poles in a similar way. However, a N and a S pole attract one another. These results may be summed up by saying:

Like poles repel, unlike poles attract.

22.1 MAKING MAGNETS

If a piece of iron is placed near a magnet it will become a magnet, but when it is removed it loses its magnetism. There are, however, several methods of making a more permanent magnet.

Magnets were originally made by stroking steel bars with a lodestone or another magnet. This was done in one of two ways, both illustrated in Fig. 22.1.

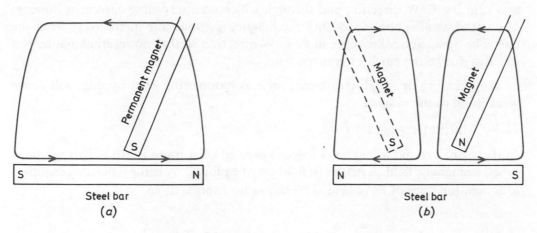

Fig. 22.1 (*a*) Single touch; (*b*) double touch

The single touch method requires one magnet (Fig. 22.1(*a*)). One pole of a magnet is stroked along a steel bar from one end to the other repeatedly. After a sufficient number of strokes the bar will begin to behave as a magnet. One of the poles produced is concentrated in one part of the bar where the stroking ends, but the other pole is apt to be spread along the bar where the stroking starts and this is a disadvantage of this method. In order to overcome the problem and concentrate the poles at both ends the method of double touch is used. The steel is stroked from the centre outwards with the unlike poles of two magnets simultaneously (Fig. 22.1(*b*)). In both methods the polarity produced at the end of the bar where stroking ceases is opposite to that of the stroking pole.

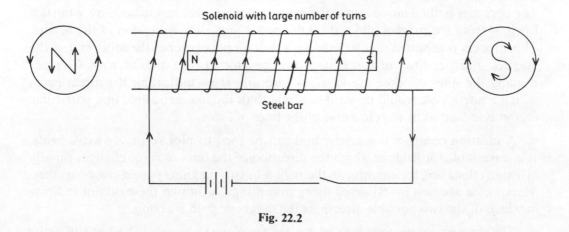

Fig. 22.2

Magnetisation by an electrical method (Fig. 22.2) is a far quicker and more efficient method of making magnets than any of those already described. A cylindrical coil is wound with about 500 turns of insulated copper wire and connected to a direct current supply. A coil of this kind is called a **solenoid**. A steel bar is placed inside the coil and the current switched on. The current creates a strong magnetic field within the coil and the steel immediately becomes a magnet, retaining its magnetism when the current is switched off.

22.2   DEMAGNETISATION

The best way to demagnetise a magnet is to withdraw it slowly from a coil whose axis is in the E–W direction and through which an alternating current is flowing. The withdrawal is continued until the magnet is about one metre away from the coil. The operation takes place in an E–W direction so that no residual magnetism remains due to the earth's magnetic field.

Any heating or rough treatment, such as hammering or dropping, will cause weakening of the magnet.

22.3   MAGNETIC FIELDS

In the space around a magnet a force is exerted on a piece of iron. This region is called a **magnetic field**. A magnetic field may be plotted by using a plotting compass to follow the lines of force. Fig. 22.3 shows how this is done.

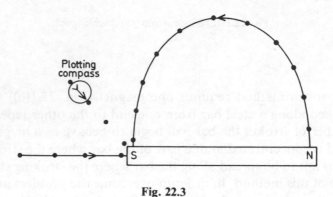

Fig. 22.3

A bar magnet is placed on a sheet of white paper. Starting near one end of the magnet, the positions of the ends of the compass needle are marked by pencil dots. The compass is then moved until the near end of the needle is exactly over the dot furthest from the magnet and a third dot made under the other end of the needle. This process is repeated many times until the compass reaches the other end of the magnet. Further lines of force may then be plotted in a similar way. Conventionally the lines of force are labelled with an arrow indicating the direction in which a north pole would move. The strength of the magnetic field in a particular region is indicated by the closeness of the lines of force.

A plotting compass is sensitive and can be used to plot relatively weak fields. It is unsuitable for fields in which the direction of the lines of force changes rapidly in a short distance, for example in the region of two magnets placed close together. These fields are best investigated using iron filings, although these do not indicate in which of the two possible directions the magnetic field is acting.

The magnets whose field is to be studied are placed beneath a sheet of stiff white paper. A thin layer of iron filings is then sprinkled from a caster. If the paper is tapped gently the filings form a pattern indicating the lines of force. Each filing becomes magnetised by induction and when the paper is tapped the filings vibrate and are able to turn in the direction of the magnetic force.

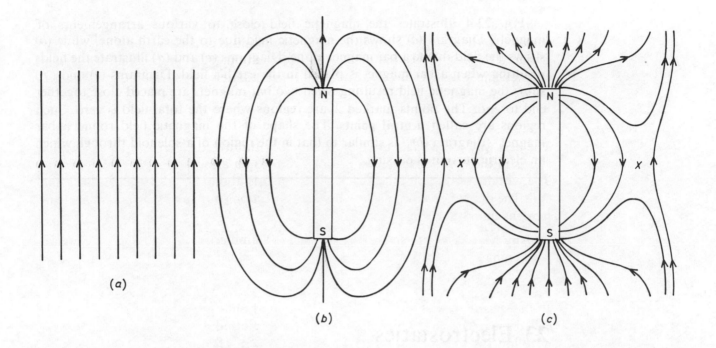

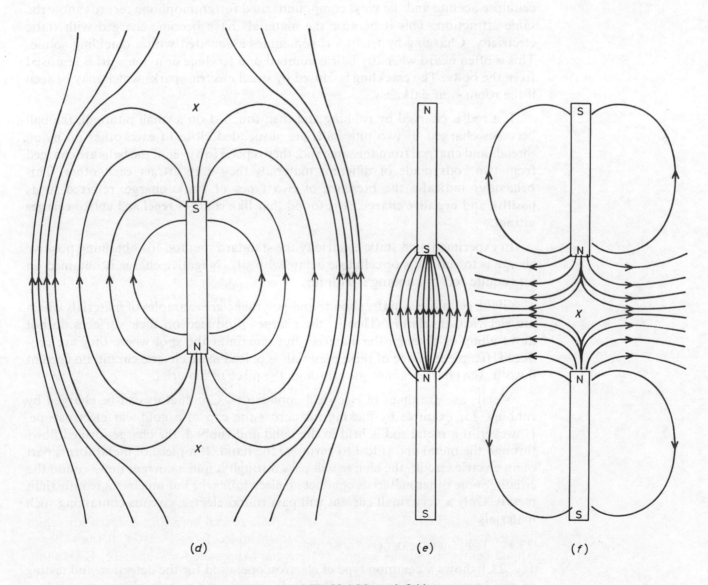

Fig. 22.4 Magnetic fields

Fig. 22.4 illustrates the magnetic field close to various arrangements of magnets. Diagram (*a*) shows the magnetic field due to the earth alone, while (*b*) shows the field due to a bar magnet alone. Diagrams (*c*) and (*d*) illustrate the fields resulting when a bar magnet is placed in the earth's field. Diagrams (*e*) and (*f*) show the magnetic field resulting when two bar magnets are placed close together end to end. The points marked *X* are regions where the total field is zero. Such regions are called **neutral points**. The shape of the magnetic field round a bar magnet – diagram (*b*) – is similar to that in the region of a solenoid through which an electric current is passing.

23 Electrostatics

If a rubber balloon is rubbed with a duster it will attract dust particles. Perspex, cellulose acetate and the vinyl compounds used for gramophone records show the same attraction. This is because the materials have become charged with static electricity. Charging by friction is sometimes associated with a crackling sound. This is often heard when dry hair is combed or a terylene or nylon shirt is removed from the body. The crackling is caused by small electric sparks which may be seen if the room is in darkness.

If a rod is charged by rubbing and then touched on a small pith ball, the ball becomes charged. If two pith balls are suspended close to each other on nylon threads and charged from the same rod, they repel. However, if the balls are charged from two rods made of different materials they may attract each other. This behaviour indicates the presence of two types of static charge, referred to as **positive** and **negative** charge. It is found that **like charges repel and unlike charges attract**.

In experiments on static electricity the standard method for obtaining positive charge is to rub glass or cellulose actate with silk. Negative charge is obtained on an ebonite rod by rubbing it with fur.

Glass, cellulose acetate, ebonite and polythene are examples of materials which are electrical insulators. That is, the charges produced on their surfaces do not move along or through the material, but remain at the spot where they are produced. If a piece of one of these materials is placed in an electric circuit, no current flows as the charge cannot pass through the piece of material.

Metals are examples of electrical conductors. Conductors can be charged by rubbing, for example by flicking fur across the cap of a gold-leaf electroscope. However, if a metal rod is held in the hand and rubbed, the charge formed flows through the metal and is lost to earth *via* the hand. If a piece of metal forms part of an electric circuit, the charge will pass through it and a current flows round the circuit. Some other materials conduct an electric charge but much less readily than metals. Only a very small current will pass round electric circuits containing such materials.

23.1 THE GOLD-LEAF ELECTROSCOPE

Fig. 23.1 shows a common type of electroscope, used for the detection and testing of small quantities of an electric charge. It consists of a metal rod surmounted by a

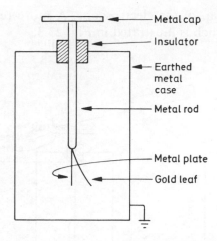

Fig. 23.1 Gold-leaf electroscope

metal cap. A metal plate with a thin gold leaf attached is fixed to the lower end of the rod. The whole apparatus is protected from draughts by housing it in an earthed metal case with glass windows.

If a rod of some suitable material is charged by friction and then brought near to the cap the leaf rises. If, for example, the rod is positively charged it will attract negative charges–from the atoms of the rod, plate and leaf–to the cap, leaving excess positive charges on the plate and leaf. The leaf and plate thus have the same charge, repulsion occurs and the leaf rises. This process is called **induction**, as the rod has not been brought into contact with any part of the gold-leaf electroscope, and no charge can therefore have been transferred from the rod to the electroscope. When the rod is removed from the cap, the leaf falls.

If the charged rod is scraped across or rolled over the surface of the cap, the leaf rises and remains up when the rod is removed. It may be necessary to repeat this procedure several times before this process of charging by contact is successful. Some of the charge on the surface of the rod has been transferred to the electroscope. If the cap of the charged electroscope is now touched with a finger, the charge flows to earth through the experimenter's body and the leaf falls. This is called **earthing**.

Fig. 23.2(*a*) illustrates the method of charging by contact and (*b*) discharging by earthing.

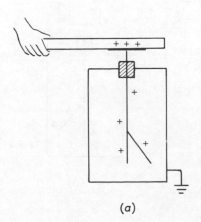

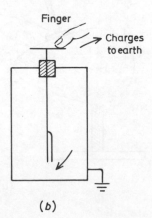

(*a*)   (*b*)

Fig. 23.2

Successful charging of an electroscope by contact is difficult as the rod is an insulator and the charges do not flow over its surface. They have to be scraped off different areas by the cap of the electroscope. A more reliable way is to use the method of induction which is illustrated in Fig. 23.3.

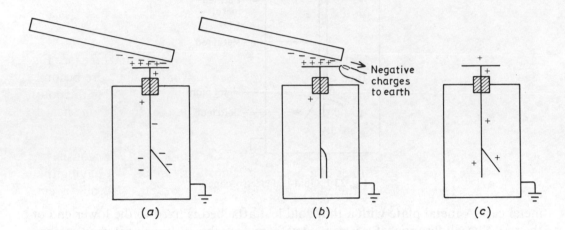

Fig. 23.3 Charging by induction

If an ebonite rod which has been charged by friction is brought close to the electroscope the leaf rises as shown in (*a*). With the rod still in place, the electroscope is earthed by touching any part of it which is conducting and the leaf falls (*b*). This happens because the negative charges previously on the plate and leaf are repelled to earth by the negative charges on the rod. However, the positive charges on the cap are still held in place by the attraction of the negative charges on the rod. The earth connection is removed. The rod is now removed and the positive charges distribute themselves throughout the electroscope (*c*); as a result the leaf rises.

23.2   DISTRIBUTION OF CHARGE OVER THE SURFACE OF A CONDUCTOR

This can be investigated using a gold-leaf electroscope and a proof plane, which consists of a small metal disc at the end of an insulating handle. The proof plane is pressed into contact with the surface of the conductor at various places in turn, and the charge acquired transferred to the electroscope (Fig. 23.4). The divergence

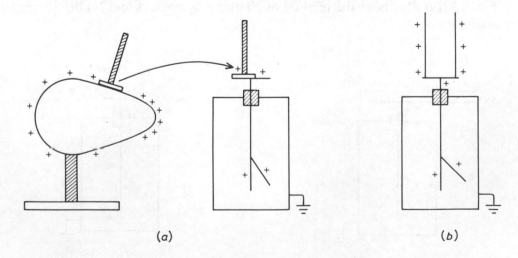

Fig. 23.4 Charge distribution

of the leaf gives a rough measure of the amount of charge transferred, and hence the surface density of the charge on the conductor. **Surface density is defined as the quantity of a charge on a unit area of a conductor.**

Fig. 23.4 shows how a charge is distributed over the surface of conductors of different shapes. It will be noticed that the charge is most densely concentrated at places where the surface is sharply curved. This is particularly noticeable at the pointed end of the pear-shaped conductor. The charge also avoids hollows as can be seen from the lack of it on the inside of the can.

23.3 POINT ACTION

When a pointed wire is connected in the dome of a Van de Graaff machine (see Fig. 23.6 and unit 23.4), the surface density of charge on the end of the wire becomes very high and a current of air, called an **electric wind**, streams away from the point. The presence of this wind can be demonstrated by bringing a candle flame near to the point. The flame is deflected by the draught.

The electric wind may also be used to work a 'windmill'. This consists of several wires arranged as spokes of a wheel on an insulated pivot and having their ends bent at right angles to their length (Fig. 23.5). When the 'windmill' is con-

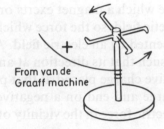

From van de Graaff machine

Fig. 23.5 'Windmill'

nected to a Van de Graaff machine an electric wind streams from the points, and the resulting reaction on the wires causes the mill to rotate in the opposite direction.

Air contains a certain number of positive and negative ions (charges). Therefore

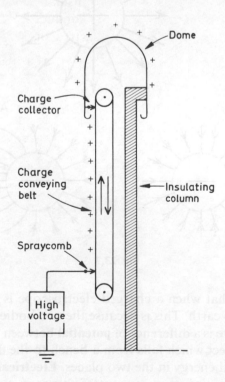

Dome

Charge collector

Charge conveying belt

Insulating column

Spraycomb

High voltage

Fig. 23.6 Van de Graaff machine

near a highly charged point conductor ions of the same sign will be strongly repelled. When these fast-moving charges collide with air molecules, further ions are formed, and in a short time an avalanche of charges will be moving rapidly away from the point. The surrounding air molecules get caught up in this stream, thus setting up the electric wind. This process is known as point action.

23.4   THE VAN DE GRAAFF MACHINE

This machine (Fig. 23.6) is a convenient way of producing a charge. A high potential is applied to a spraycomb consisting of a series of sharp points adjacent to a long moving belt made of synthetic insulating material. An electric charge is sprayed from the comb to the belt by point action and is carried up to the top of an insulating column. Here the charge is removed from the belt by another spraycomb connected to the inside of the dome. Thence the charge passes to the outside surface of the dome and cumulatively builds up a very high potential on it. The value of the potential is limited by the breakdown voltage of the surrounding air.

23.5   LINES OF FORCE

In the same way that the force which a magnet exerts on a nearby compass needle can be represented by a magnetic field, so the force which a charged body exerts on a nearby charge can be represented by an electric field. An **electric line of force** is a line drawn in an electric field such that its direction at any point gives the direction of the electric force on a positive charge placed at that point. Thus lines of electric force begin on a positive charge and end on a negative charge of the same size. Fig. 23.7 shows the lines of electric force in the vicinity of different arrangements of electric charges.

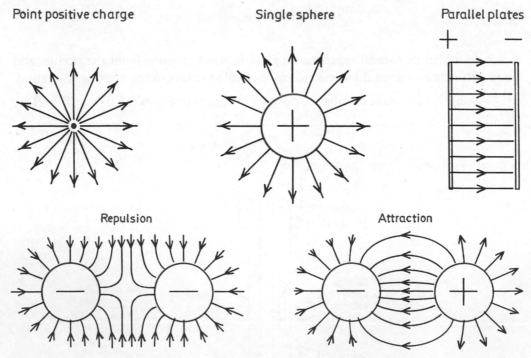

Fig. 23.7

23.6   POTENTIAL

We have already seen that when a charged electroscope is earthed, charges flow from the electroscope to earth. This is because the two bodies have different values of potential; that is, there is a difference of **potential** between them. The situation is similar to that of an object which falls from a bench to the floor because there is a difference in its potential energy in the two places. **Electrical potential difference** is what determines the direction of movement of charge in a similar way to the difference in potential energy determining the direction of movement of an object.

When making measurements of electric potential it is necessary to choose a convenient zero, as it is with the potential energy of an object. In this case we take the earth as the zero of electric potential, and all other values are thus relative to it.

So far we have considered the electroscope as an instrument for indicating the presence of a small quantity of charge. In fact the divergence of the leaf of an electroscope indicates that there is a potential difference between the leaf and the case as can be shown by considering the simple experiment shown in Fig. 23.8.

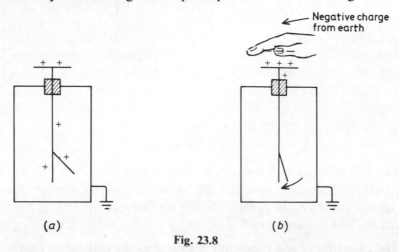

Fig. 23.8

In diagram (*a*) the electroscope is shown with a small positive charge. In diagram (*b*) the experimenter's hand has been brought near to the cap and the leaf has fallen to some extent. This is because the potential of the electroscope has been lowered. The explanation is as follows. The positive charge on the cap attracts a negative charge from the earth to the hand when the hand is brought close to the cap. To remove the hand energy will have to be expended to overcome the attraction of the charges, thus increasing the potential energy of the electroscope. Thus when the hand has been removed the potential energy of the electroscope is greater than when the hand is present. As the hand is taken away the leaf rises to its former height. The fall and rise in the leaf cannot be due to a change in the charge on the electroscope as no contact is made with the electroscope throughout the experiment. The movement of the leaf must therefore indicate the change in potential energy of the electroscope.

The potential is the same over the entire surface of a charged conductor, no matter how the charges may be distributed. In fact the charges take up the particular distribution over the surface so that the potential is the same everywhere. Suppose there were a difference in potential between two points on the surface of a conductor. This potential difference would cause charge on the surface to move until the difference no longer existed.

23.7 CAPACITANCE

A conductor with a large **capacitance** has the ability to store more charge at a given potential than one with a smaller capacitance.

If two equal quantities of charge (*Q* units) are given to conductors of different sizes they will acquire different potentials. This may be shown by standing two unequal metal cans on the caps of two similar electroscopes (Fig. 23.9(*b*) and (*c*)).

The cans are given equal charges using a metal disc on an insulating handle and an ebonite slab is charged by rubbing. The disc is touched on the ebonite and then lowered into one of the cans (Fig. 23.9(*a*)). If the disc is moved about well inside the can the divergence of the leaf remains constant. This means that all the field lines, from the charges on the disc, end on the inside of the can and, if the can is now touched by the disc on the inside of its base, a complete transfer of the charge to

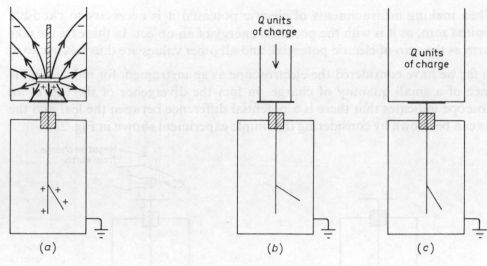

Fig. 23.9

the can will take place. If the process is now repeated for the other can, one may be sure that each can has received the same quantity of charge. However, the leaf divergence is seen to be greater for the smaller can indicating that it is at a higher potential than the larger can. The larger can is said to have a bigger capacitance.

The capacitance of a conductor is defined as the ratio of its charge to its potential.

$$\text{Capacitance} = \frac{\text{charge}}{\text{potential}}.$$

An electric charge is measured in coulombs, potential in volts, and capacitance in farads.

The unit of capacitance, the **farad (F)**, is defined as **the capacitance of a conductor such that a charge of 1 coulomb changes its potential by 1 volt.**

In practice capacitance is often measured in microfarads (μF). A microfarad is a millionth part of a farad.

23.8   Capacitors

A capacitor is a device for storing a charge. The simplest example is the parallel plate capacitor. It consists of two flat metal plates placed a small distance apart. The space between the plates is often filled by a sheet of insulating material, such as polythene. The three factors which affect the size of the capacitance of a parallel plate capacitor are the area of the plates, their separation and the nature of the insulating material (dielectric) placed between them. Increasing the area of the plates increases the capacity of the system as more charge can be accommodated without increasing the charge density and hence the potential difference of the plates. Increasing the separation of the plates with a given charge requires work to be done and hence increases the potential of the capacitor. It is then storing the same charge at a higher potential. From the definition of capacitance in the last section it will be seen that it now has a lower value of capacitance. Placing an insulator between the plates immediately lowers the potential difference between them. It can be seen from the same definition that this leads to an increase in the value of their capacitance.

All these factors can be summed up in the equation:

$$C = \frac{\varepsilon_0 A}{d}$$

where A is the area of the plate, d their separation, and ε_0 depends on the material filling the space between them.

Capacitors are important components of radio circuits. Fig. 23.10 illustrates three common types.

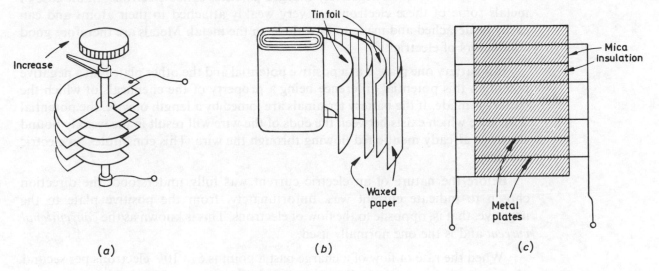

Fig. 23.10 Capacitors

A variable capacitor is used for tuning radios; it is shown in diagram (*a*). One set of plates is fixed and the other is rotated by a knob to alter the area of the plates which overlap. Diagram (*b*) illustrates the construction of a paper capacitor. This type is cheap but only useful at low frequencies. It contains two long strips of tinfoil separated by thin waxed paper. The whole is rolled up and sealed in a metal box to exclude moisture. Another type which may be used in a radio receiver is a fixed capacitor having mica as the medium between the plates as shown in diagram (*c*). Two sets of metal plates are each connected to a terminal. Mica capacitors may range from small values such as 0.0001 μF to higher values such as 1 μF. For large values of capacitance electrolytic capacitors are often used.

23.9 SUMMARY

1. Like electrical charges repel, unlike charges attract.
2. An ebonite rod rubbed with fur, or a rubbed polythene strip, acquires a negative charge; a glass rod rubbed with silk or a rubbed cellulose acetate strip acquires a positive charge.
3. Metals, the human body and the earth are examples of conductors; glass, ebonite and plastics are normally insulators.
4. A conductor can be charged by contact or induction.
5. An induced charge can be obtained on an insulated conductor by: (i) bringing a charge near; (ii) touching the conductor; (iii) removing the finger; (iv) taking the charge away.
6. A pointed charged conductor has a high density of charge at the point. This is shown by the rotation of a metal windmill.
7. Electric fields exist all round conductors which carry electric charges.
8. Positive charges move from one point to another at a lower potential when they are joined by a conductor.
9. The capacitance C of a capacitor in farads $= Q/V$, where Q is the charge in coulombs, and V is the potential difference in volts.
10. The capacitance of a capacitor increases: (i) the greater the common area of the plates; (ii) the smaller the distance between the plates; (iii) the higher the constant of the dielectric medium between the plates.

24 Current electricity

All atoms possess small negatively charged particles called electrons. In the case of metals some of these electrons are very weakly attached to their atoms and can easily be detached and made to flow through the metal. Metals are therefore good conductors of electricity.

In a battery one plate is at a positive potential and the other plate is at a negative potential, this potential difference being a property of the chemicals of which the battery is made. If the battery terminals are joined by a length of wire, the potential difference which exists between the ends of the wire will result in the weakly bound electrons already mentioned flowing through the wire. This constitutes an electric current.

Before the nature of an electric current was fully understood, the direction chosen to indicate current was, unfortunately, from the positive plate to the negative, that is, opposite to the flow of electrons. This is known as the *conventional current* and is the one normally used.

When the rate of flow of a charge past a point is 6×10^{18} electrons per second, the current is 1 **ampere (A)**. The formal definition is given in unit 25.4.

As an electric current is a flow of an electric charge, the quantity of an electric charge which passes any point in a circuit will depend on the strength of the current and the time for which it flows. The unit of electric charge is the **coulomb. A coulomb is the quantity of an electric charge conveyed in one second by a steady current of one ampere.**

Thus   **charge (Q) = current (I) × time (t)**
$$Q = It.$$

24.1   POTENTIAL DIFFERENCE

In order to achieve current flow in a circuit a potential difference V must exist. The unit of potential difference is the **volt. Two points are at a potential difference of 1 volt if 1 joule of work is done per coulomb of electricity passing between the points.**

24.2   ELECTROMOTIVE FORCE

The maximum potential difference that a cell is capable of producing in a circuit is called its **electromotive force (e.m.f.)**. It is measured in volts and may be regarded as the sum total of the potential differences which it can produce across all the components of a circuit in which it is connected, including the potential difference required to drive the current through itself. The e.m.f. is a property of the materials in the cell and is not affected by the nature of the circuit into which the cell is connected. On the other hand the potential difference which the cell produces in the external circuit connected to it depends on the current flowing through the circuit.

The **e.m.f. of a cell in volts is defined as the total work done in joules per coulomb of electricity conveyed in a circuit of which the cell is a part.**

24.3   RESISTANCE

As the potential difference between the ends of a conductor is increased the current passing through it increases. **If the temperature of the conductor does not alter, the current which flows is proportional to the potential difference applied (Ohm's law).** Fig. 24.1 shows this effect.

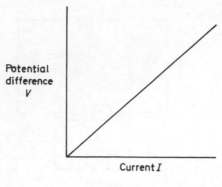

Fig. 24.1

The gradient of this graph has a constant value, obtained by dividing the potential difference at any point by the current. The value of this constant gradient is known as the **resistance R** of the conductor. The unit of resistance is the **ohm (Ω)**.

$$\frac{V}{I} = \text{constant } (R) \text{ (Ohm's law)},$$

hence $\qquad\qquad V = IR.$

A good conductor is one with a low resistance, a poor one has a high resistance.

In some conductors the current is not proportional to the potential difference between its ends. In a light bulb V/I is found to increase with temperature, that is R is not constant but increases as the temperature of the filament of the bulb increases. Fig. 24.2(a) shows the relationship between potential difference and current in such cases.

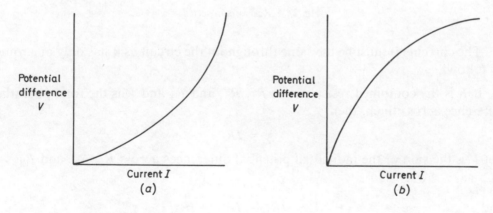

Fig. 24.2

Fig. 24.2(b) shows results that might be obtained from a thermistor for which the resistance decreases with increasing temperature.

24.4 AMMETERS AND VOLTMETERS

An **ammeter** is used to measure the electric current through a conductor and must be in series with the conductor. It has a resistance; however, this should be as small as possible so that it only reduces the current to be measured by a very small amount.

A **voltmeter** measures potential difference between two points in a circuit. It must be in parallel with the part of the circuit concerned. In order that it should take as little current as possible out of the main circuit it must have as large a resistance as possible.

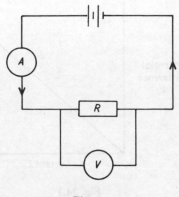

Fig. 24.3

Fig. 24.3 shows an ammeter Ⓐ and a voltmeter Ⓥ correctly connected to determine the value of the resistance *R*.

24.5   RESISTORS IN SERIES

A number of **resistors** R_1, R_2, R_3, are said to be connected **in series** if they are connected end to end as shown in Fig. 24.4.

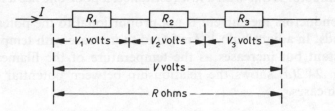

Fig. 24.4 Resistors in series

The current (*I*) must be the same throughout the circuit as it has only one route to follow.

If *R* is the combined resistance of R_1, R_2, and R_3 and *V* is the total potential difference across them, then

$$V = IR$$

but *V* is the sum of the individual potential differences across R_1, R_2, and R_3.

Thus $\qquad\qquad\qquad\qquad V = V_1 + V_2 + V_3$

and $\qquad\qquad\qquad\qquad V = IR_1 + IR_2 + IR_3,$

therefore $\qquad\qquad\qquad IR = IR_1 + IR_2 + IR_3$

and hence $\qquad\qquad\qquad \mathbf{R = R_1 + R_2 + R_3.}$

The same argument may be applied to any number of resistors in series.

In Fig. 24.3 the electromotive force *E* of the battery must not only provide the potential difference to drive the current through the resistors, but also through itself V_b. The battery is said to have its own resistance *r*, known as its **internal resistance**. The total resistance of the circuit must include *r*, and is equal to the sum of *r*, R_1, R_2 and R_3.

Now $\qquad\qquad\qquad\qquad V = V_1 + V_2 + V_3$

and $\qquad\qquad\qquad\qquad E = V_1 + V_2 + V_3 + V_b,$

hence $\qquad\qquad\qquad\qquad E - V = V_b.$

V_b is usually referred to as the *lost volts* in the circuit.

24.6 Resistors in parallel

Resistors are said to be **in parallel** when they are placed side by side in a circuit (Fig. 24.5).

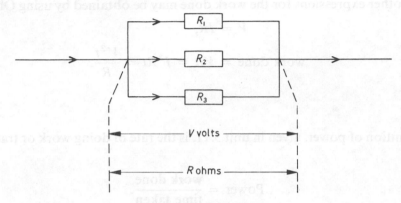

Fig. 24.5 Resistors in parallel

The conductance of each branch of the circuit is I/V, that is the quantity of current which passes for each volt of potential difference applied. But I/V is equal to $1/R$; thus the conductance of each branch of a circuit is the reciprocal of its resistance. The total conductance of the three resistors in parallel is clearly the sum of the conductance of each individual resistor, as each one provides the current in the circuit with an alternative route, and thus makes its passage easier.

Thus
$$\frac{1}{R} = \frac{1}{R_1} + \frac{1}{R_2} + \frac{1}{R_3}.$$

The same argument may be applied to any number of resistors in parallel. As an example consider a 2-ohm resistor and a 4-ohm resistor in parallel with each other.

$$\frac{1}{R} = \frac{1}{2} + \frac{1}{4} = \frac{3}{4}.$$

Thus
$$R = \tfrac{4}{3} = 1\tfrac{1}{3} \text{ ohms.}$$

The combined resistance of any number of resistors in parallel is always less than the value of any one of them. This is clear when one realises that placing one resistor in parallel with another provides an alternative route for the current and thus eases its passage round the circuit.

24.7 Energy

When a potential difference is applied to the ends of a conductor some of the electrons inside it are set in motion by electric forces. Work is therefore done and the electrons acquire energy. The moving electrons form an electric current, and the energy of this current appears in various forms according to the type of circuit of which the conductor forms a part. For example in an electric fire the energy of the current is largely made available as heat, in an electric light as heat and light, and in an electric motor as mechanical energy of rotation. Energy in these various forms is produced at the expense of the source of electricity.

From the definition of the volt it can be seen that if a potential difference of one volt is applied to the ends of a conductor and one coulomb of electricity passes through it, then the work done, or energy transformed, is 1 joule. Hence if the potential difference applied is V volts and the quantity of electricity that passes is Q coulombs then the work done is QV joules.

However,   Charge (Q) = current (I) × time (t)

or $Q = It$,

therefore   work done = $QV = VIt$ joules.

Two other expressions for the work done may be obtained by using Ohm's law.

As $V = IR$,

$$\text{work done} = VIt = I^2Rt = \frac{V^2t}{R}.$$

The definition of power, given in unit 3.11, is the rate of doing work or transferring energy.

$$\text{Power} = \frac{\text{work done}}{\text{time taken}}.$$

If the work done is measured in joules, the unit of power is the **watt**. Using the equations for energy given in the last unit, power may be obtained by dividing by time.

$$\text{Power} = VI = I^2R = \frac{V^2}{R} \text{ watts.}$$

The first expression in words is:

$$\text{Power} = \text{potential difference} \times \text{current.}$$

Example:

Find (a) the current taken by, and (b) the resistance of, the filament of a lamp rated at 240 volts, 40 watts.

$$\text{Power} = \text{potential difference} \times \text{current,}$$

therefore $\text{current} = \dfrac{\text{power}}{\text{potential difference}}$

$$= \frac{40}{240} = \frac{1}{6} = 0.16 \text{A};$$

but   potential difference = current × resistance,

thus $\text{resistance} = \dfrac{\text{potential difference}}{\text{current}}$

$$= \frac{240}{1/6} = 1440 \text{ ohms.}$$

24.9   Cost

If an electricity meter is inspected it will be found to have the abbreviation 1 kWh on it. This stands for kilowatt-hour, the commercial unit of electrical energy. It is the energy supplied in one hour by a rate of working of 1000 watts.

1 kilowatt-hour = 1000 watt-hours
= 1000 joules/second for 1 hour
= 1000 × 60 × 60 joules
= 3 600 000 joules or 3.6 MJ.

Example: Find the cost of running a 2 kW fire for 3 hours if the cost of electrical energy is 4.0 pence per kilowatt-hour (unit).

$$\text{Total energy consumed} = \text{power} \times \text{time}$$
$$= 2 \times 3 \text{ kWh}$$
$$= 6 \text{ kWh}.$$
$$\text{Cost} = 6 \times 4.0 = 24 \text{ pence}.$$

24.10 HOUSE ELECTRICAL INSTALLATION

The cable bringing the mains electricity supply into a house contains two wires, one of which is 'live', and the other 'neutral'. The neutral wire is earthed at the local transformer substation, so it is at earth potential. At some convenient place inside the house the mains cable enters a sealed box, where the live wire is connected to the Electricity Board's fuse. On the far side of this box the power cable enters the meter and from there it goes to the main fuse box. The fuse box contains a separate fuse for each of the lighting circuits, ring circuit and cooker circuit (Fig. 24.6).

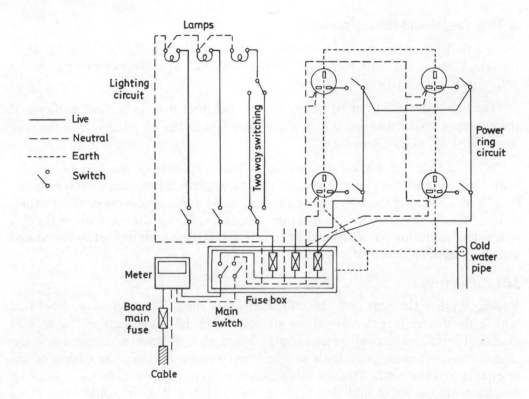

Fig. 24.6 House electrical installations

In modern installations the power sockets are tapped off a ring circuit. This cable passes through the various rooms in the house and has both its ends connected to the mains supply. Thus there are two paths by which the current may get to a particular socket, which effectively doubles the capacity of the cable.

It can be seen from Fig. 24.6 that all light and power switches and fuses are placed in the live side of the supply. If they were in the neutral side all light and power sockets would remain live when the switches were in the off position.

24.11 FUSES

For safety reasons all domestic electrical appliances are fused. That is, in at least one place on each circuit, a fuse is fitted, which will 'blow' if a fault develops in the circuit and too great a current passes through it. This arrangement protects the wiring against the possibility of overheating and setting fire to the house.

The main house fuses for the different circuits generally consist of short lengths of tinned copper wire fitted into porcelain carriers. These fuses are usually situated under the stairs or in a cupboard which also contains the electricity meter and Electricity Board fuse.

In addition, items which are separately plugged into power sockets–for example, fires–contain a fuse in the plug. This type of fuse consists of a small glass cartridge with a thin wire through its centre. Such a fuse is rated at 2, 5, 10 or 13 amps to suit the appliance to which it is connected. For example, the value of the fuse which should be used in the plug of a 240 V, 2 kW fire is calculated as follows.

$$\text{Power} = \text{voltage} \times \text{current},$$

thus

$$2000 = 240 \times \text{current}$$

and the

$$\text{current} = \frac{2000}{240} \text{ A} = 8.33 \text{ A}.$$

A 10 A fuse would thus be suitable.

If a fault develops in the appliance, this fuse will normally 'blow' rather than the larger one in the main fuse box. This avoids other appliances being put out of action at the same time.

The thickness of wire used in any fuse is such that it will overheat and melt if the current passing through it exceeds its specified rating by much. Once the wire has melted the circuit is broken.

When a fuse has 'blown' it is essential that the cause of the fuse blowing is found before it is repaired or replaced and the appliance used again. Sometimes a fuse 'blows' because the fuse wire is very old and has become weakened by oxidation, or it can blow due to a fault in the circuit, such as a short circuit in the flex where the insulation has worn and frayed. Whatever the fault it must be found and put right before a new fuse is fitted.

24.12 EARTHING

Besides the live (brown) and neutral (blue) wires, all power circuits are provided with a third wire (green and yellow stripes) which has been earthed by a good electrical joint to the cold water supply. When an appliance is connected to the circuit the earth wire provides a low resistance route between the casing of the appliance and the earth. This is a safeguard to prevent anyone receiving a shock by touching the casing should this become 'live'. Such a danger would arise if the insulation on the live flex had become worn and allowed the live wire to come into contact with the casing of the appliance. If this happened in a properly earthed appliance, a large current would instantaneously flow to earth and the fuse would blow, thereby cutting off the supply.

Loose connections in plugs are another potential source of danger to the person. Proper earthing again removes the danger.

25 Electromagnetism

Whenever an electric current flows in a conductor a magnetic effect is present in the region of the conductor. This can most easily be demonstrated by passing a current down a vertical conductor, such as a retort stand (Fig. 25.1).

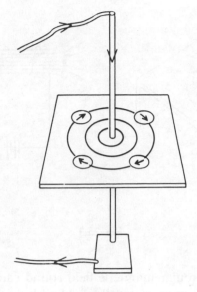

Fig. 25.1

At some convenient height the conductor passes through a hole in a horizontal platform. A number of small compasses are placed on the platform. When the current is switched on the compass needles swing from pointing to magnetic north and re-align themselves in a circular path around the wire. The influence of the current in the region of the wire is called a **magnetic field**.

This result can be remembered using the following rule. Imagine the wire to be grasped in the right hand with the thumb pointing along the wire in the direction of the current. The direction of the fingers will give the direction of the lines of force. This rule is purely an aid to memory; it does not explain how the lines occur.

A knowledge of the magnetic field around a straight conductor can be used to predict the pattern round other simple geometrical arrangements of current carrying wires–for example, a flat coil and a solenoid.

The pattern of lines of force when a current is passed through a flat circular coil is shown in Fig. 25.2.

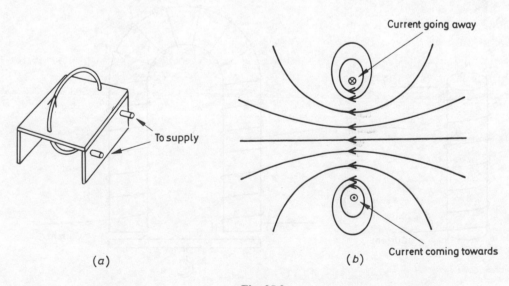

(a)   (b)

Fig. 25.2

Fig. 25.3 illustrates the magnetic field due to a current in a solenoid.

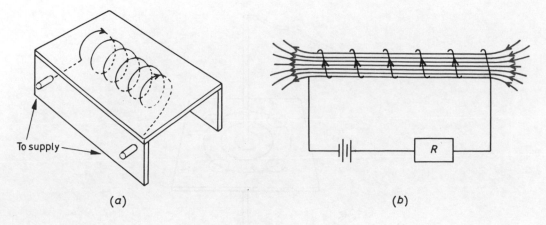

Fig. 25.3

If one considers the circular magnetic field round each short length of wire in the flat circular coil, it can be seen that the field adds up through the centre of the coil. This leads to a strong field through the coil but a weak one outside it (Fig. 25.2(*b*)). A solenoid may be considered as a series of flat circular coils, each a little spaced from one another on a common axis. Each turn of insulated wire gives a magnetic field similar to that of a flat circular coil. The fields between neighbouring turns oppose one another and cancel, but the fields along the common axis reinforce, producing the pattern shown in Fig. 25.3(*b*).

The magnetic fields resulting from a current passing through flat coils and solenoids of square cross-sectional area are similar to those shown for circular areas.

25.1   THE ELECTROMAGNET

The use of a solenoid for making magnets has already been described in unit 22.1. When a piece of steel is placed inside a solenoid and the current switched on, the steel becomes a magnet and remains one when the current is switched off. If a piece of soft-iron is used instead, the iron acts as a magnet only while the current is switched on. Fig. 25.4(*a*) shows such an electromagnet. Sometimes the cores of

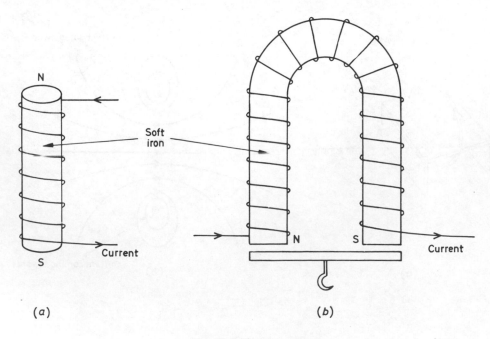

Fig. 25.4 Electromagnet

electromagnets are U-shaped (Fig. 25.4(*b*)). This arrangement has the advantage that the attraction of both ends can be used simultaneously.

Electromagnets are very strong. A small one, made in a laboratory from a U-shaped core and a few turns of wire each carrying a current of a few amps, is capable of lifting several kilograms. Those used in industry can lift several tonnes, for example car and lorry bodies. Yet when the current is switched off, the magnetism ceases and the load is released.

25.2   THE ELECTRIC BELL

The electric bell consists of two solenoids wound in opposite directions on two soft-iron cores joined by a soft-iron yoke (Fig. 25.5). One end of the windings is

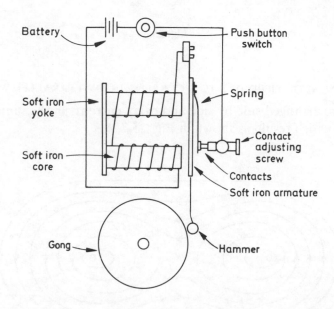

Fig. 25.5 Bell

connected to the battery and the other to a metal bracket which supports a spring-mounted soft-iron armature (moving part). The armature carries a light spring to which is soldered a small platinum disc as contact. The disc presses against the end of a platinum-tipped contact screw from which a wire goes to the push switch.

When the switch is pressed a current flows through the circuit and the cores become magnetised. The resultant attraction of the armature separates the contacts and breaks the circuit. The cores become demagnetised and the armature is returned to its original position by the spring. Contact is re-made and the action repeated. As a result the armature vibrates and a hammer attached to it strikes the gong.

25.3   MOVING-IRON INSTRUMENTS

The attraction type of moving-iron instrument is shown in Fig. 25.6. When a current is passed through the solenoid it becomes magnetised and a specially shaped piece of soft-iron is attracted into it with a force which depends approximately on the square of the current. The soft-iron pivots and a needle moves across the calibrated scale. Control is provided by a hairspring.

Moving-iron instruments have two main advantages. Firstly they can be used with alternating as well as direct current and secondly they are robust. They may, for example, be used in cars.

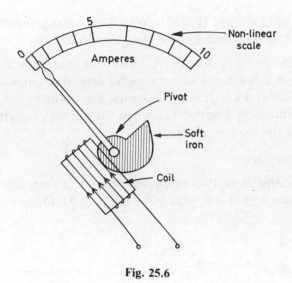

Fig. 25.6

25.4 THE MAGNETIC FIELD DUE TO A CURRENT IN TWO PARALLEL WIRES

If two wires are arranged side by side and carry currents in the same direction the magnetic field which results is shown in Fig. 25.7.

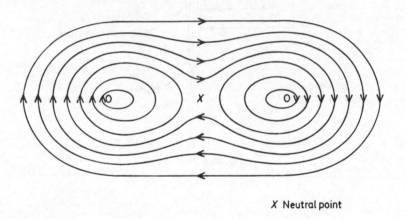

X Neutral point

Fig. 25.7

Between the wires the magnetic fields are in opposition and the resultant field is weak. At some point between the wires the magnetic fields are equal in strength and opposite in direction and the resultant field is zero. This point is known as a neutral point, and will be midway between the wires if they carry equal currents. Beyond the wires a relatively strong field results and there is a force tending to pull the wires inward.

The definition of the ampere mentioned in unit 24 comes from such an arrangement. **The ampere is the current which, if flowing in two straight parallel wires of infinite length, placed one metre apart in a vacuum, will produce between the wires a force of 2×10^{-7} newtons per metre of their length.**

If the currents in the two wires flow in opposite directions then the resultant force tends to push the wires apart.

25.5 THE FORCE ON CHARGES MOVING IN A MAGNETIC FIELD

Fig. 25.8 shows a current carrying wire placed in a magnetic field. As soon as the current is switched on the length of wire AB moves horizontally on the other two pieces. The interaction of the magnetic field due to the current flow and that due to

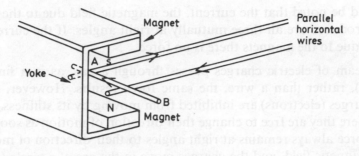

Fig. 25.8

the magnets, results in a horizontal force. The effect may be understood by reference to Fig. 25.9.

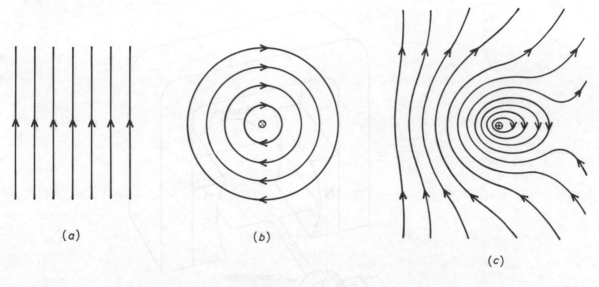

Fig. 25.9

The two fields shown separately in (*a*) and (*b*) reinforce to the left of the wire whereas they tend to cancel to the right (*c*). A strong magnetic field results on the left with the lines in tension. The tension results in a force from left to right. If either magnetic field is reversed the force reverses.

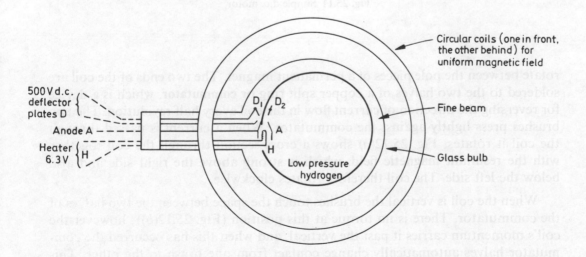

Fig. 25.10 Fine beam tube

It should be noted that the current, the magnetic field due to the magnets and the force produced are all three mutually at right angles. If the current is parallel to the field due to the magnets there is no force.

If a stream of electric charges passes through a gas, as in a fine beam tube (Fig. 25.10), rather than a wire, the same force results. However, whereas in a wire the charges (electrons) are inhibited from moving by its stiffness, this is not so in a gas, where they are free to change their direction of motion as soon as the force acts. The force always remains at right angles to their direction of motion, as well as to the magnetic field, and the charges move in the arc of a circle (see unit 3.4).

25.6   THE D.C. ELECTRIC MOTOR

Fig. 25.11 shows the construction of a simple direct current (d.c.) electric motor. It consists of a rectangular coil of wire mounted on a spindle so that it is free to

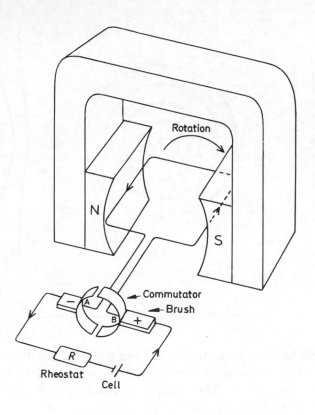

Fig. 25.11 Simple d.c. motor

rotate between the pole pieces of a permanent magnet. The two ends of the coil are soldered to the two halves of a copper split ring or **commutator**, which is a device for reversing the direction of current flow in the coil every half revolution. The two brushes press lightly against the commutator. When a current is passed through the coil it rotates. Fig. 25.12(*a*) shows a cross-section through the coil together with the resultant magnetic field, which is strong above the right side and also below the left side. The coil therefore rotates clockwise.

When the coil is vertical the brushes touch the space between the two halves of the commutator. There is no torque at this position (Fig. 25.12(*b*)); however the coil's momentum carries it past the vertical, and when this has occurred the commutator halves automatically change contact from one brush to the other. This reverses the current through the coil (Fig. 25.12(*c*)) and thus it rotates through the

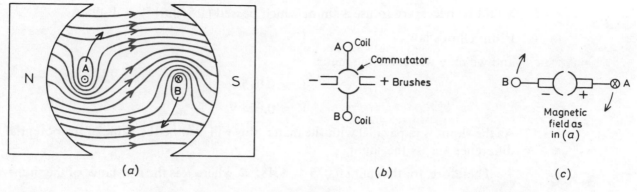

Fig. 25.12

next half turn. The reversal of the direction of current flow by the commutator each half turn ensures the continued rotation of the coil.

The simple motor described is not very efficient or powerful, as the torque changes from a maximum when the coil is horizontal to zero when it is vertical. The motor can be improved by winding a number of coils, each of many turns, at different angles round a soft-iron armature. The greater number of turns in a coil gives a greater torque; having a number of coils ranged at angles round the armature means that when one is vertical another is horizontal thus resulting in an even torque. The iron armature becomes magnetised and increases the magnetic field through the coils, resulting in a greater torque. It is also usual for the magnetic field to be provided by electromagnets rather than permanent magnets.

25.7 THE MOVING COIL METER (GALVANOMETER)

The moving coil meter relies on the same principle for its operation as the simple electric motor just described. The coil is normally mounted vertically on a phosphor bronze suspension strip or spring rather than horizontally on a spindle. A loosely coiled spring below the coil controls how far it turns; the greater the current through the coil the greater the angle through which it turns before the torque exerted by the spring equals the torque due to the magnetic field. The reading of the meter is shown by attaching a pointer to the coil and allowing it to move over a scale.

25.8 AMMETERS AND VOLTMETERS

Moving coil meters are made to take currents of only a few milliamps at the most. When larger currents have to be measured a low resistance (shunt) is placed in *parallel* with the meter (Fig. 25.13(*a*)). The greater part of the current passes through the shunt and only a small known fraction through the meter.

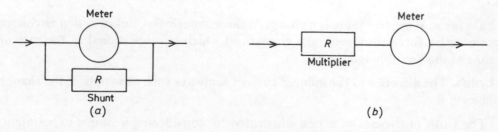

Fig. 25.13

Suppose, for example, a meter of resistance 5 ohms, capable of passing a maximum current of 15 mA is to be used for measuring currents up to 1.5 A. It would be necessary to use a shunt which passed $(1.5–0.015) = 1.485$ A.

From Ohm's law $\qquad\qquad V = IR$

and we may write, for the meter

$$V = 0.015 \times 5$$
$$V = 0.075 \text{ V}.$$

As the shunt is in parallel with the meter, this will also be the value of the potential difference across the shunt.

Therefore, for the shunt $0.075 = 1.485\,R$, where R is the resistance of the shunt.

Thus $\qquad\qquad R = \dfrac{0.075}{1.485} = 0.0505 \text{ ohm}.$

The shunt will have different values for other current ranges. In every case the total resistance of the ammeter is less than the meter alone (less than 5 ohms in this case) and the ammeter only slightly impedes the flow of current in the circuit.

A moving coil meter may be adapted for use as a voltmeter by connecting a high resistance in *series* with it (multiplier), as shown in Fig. 25.13(*b*). Suppose the meter described above is required to measure a potential difference up to 5 volts. The potential difference across the meter must not exceed 0.075 V. The p.d. across the multiplier alone is therefore equal to $(5 - 0.075) = 4.925$ V. Furthermore the current through the multiplier will be the same as that through the meter, that is 0.015 A.

Now $\qquad\qquad V = IR$

and thus, for the multiplier, $4.925 = 0.015\,R$ where R is its resistance.

Therefore $\qquad\qquad R = \dfrac{4.925}{0.015} = 328.3 \text{ ohms}.$

For all ranges the resistance of the voltmeter is relatively high. The higher its value the better as it will draw less current away from the main circuit.

26 Electromagnetic induction

The magnetic field in a region is often referred to as **magnetic flux**.

26.1 Laws

1. Faraday's. Whenever there is a change in the magnetic flux linked with a circuit an electromotive force is induced, the strength of which is proportional to the rate of change of the flux through the circuit.

2. Lenz's. The direction of the induced current is always such as to oppose the change producing it.

The truth of these laws is best illustrated by considering a simple experiment. First a centre-zero meter is connected in series with a cell and a suitable high resistance, and the direction of movement of the pointer noted when a small current passes in a known direction. The meter is now connected to the ends of a straight wire placed at right angles to the lines of magnetic field between two opposite magnetic poles (Fig. 26.1).

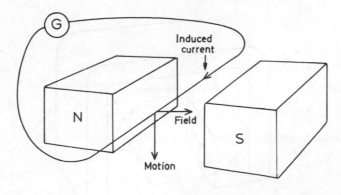

Fig. 26.1

If the wire is moved downwards the meter indicates that an induced current flows in the direction shown. If the wire is moved up the current flows in the opposite direction. The same results are obtained if the wire is held still and the magnets moved up or down respectively. It is the relative motion between the wire and magnetic field which leads to the induced voltage and hence current. The quicker the relative motion takes place the greater the deflection (Faraday's law).

While the induced current flows there is a magnetic field around the wire. Consideration of the effect of this field and that of the magnets shows that it results in an upward force in Fig. 26.1. That is, the induced current leads to a force acting in the opposite direction to the movement of the wire and thus opposing it (Lenz's law).

26.2   THE SIMPLE D.C. DYNAMO

The simple direct current electric motor described in unit 25.6 may be used as a simple d.c. dynamo. The motor is connected in series with a resistance and a moving coil meter instead of a voltage source. When the coil is rotated the meter is seen to deflect in one direction, although the deflection is not a steady one. The commutator ensures that, although the current in the coil itself reverses during the second half of a rotation, the same brush always remains positive and the other negative.

The simple dynamo just described is not very efficient. A practical dynamo has a number of coils wound in slots cut in the armature. Each coil has its own pair of segments in a multi-segment commutator. This arrangement ensures that the e.m.f. obtained is fairly steady, as it is only the horizontal coils which are connected to the brushes at any given moment. The iron armature is built in layers, each one insulated from its neighbours. Although e.m.f.'s are induced in these layers of iron as they rotate in the magnetic field, very little current flows in the armature as a result, due to this insulation.

26.3   THE SIMPLE A.C. DYNAMO

This differs from the simple d.c. version, described in the last unit, only in its connections (Fig. 26.2). The ends of the coil are connected to two slip rings mounted on the coil spindle. One side of the coil is thus always connected to the same brush.

The outputs of both a simple a.c. and a simple d.c. dynamo are shown in Figure 26.3(*a*) and (*b*) respectively. In each case the coil starts from the vertical.

26.4   ALTERNATING CURRENT

Fig. 26.3(*a*) shows how the voltage from an a.c. dynamo changes with time. One revolution of the dynamo produced one complete wave or one cycle as shown

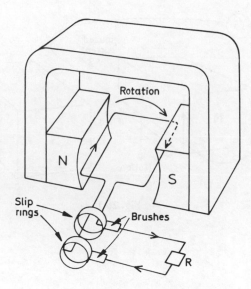

Fig. 26.2 Simple a.c. dynamo

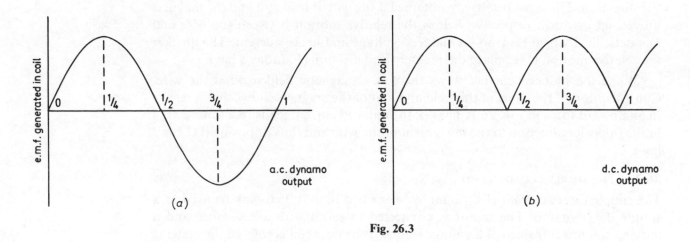

Fig. 26.3

in the diagram. The number of complete waves produced every second is known
as the frequency of the alternating current. The unit of frequency is the **Hertz (Hz)**.
In Great Britain the frequency of the mains supply is 50 Hz.

The maximum value of an alternating voltage during its cycle is known as its
peak value. However, for much of the time, the magnitude of the voltage is considerably less than this, and an alternating voltage of a particular peak value will
supply energy at a slower rate than a steady voltage of the same value. It is thus
usual to express the root-mean-square (r.m.s.) value of the voltage which is related
to the peak voltage by the equation

$$E_{\text{r.m.s.}} = \frac{E_p}{\sqrt{2}}.$$

The **root-mean-square voltage** is the steady voltage that would produce the same
rate of energy supply as the alternating voltage of peak value E_p. In Great Britain
the 240 V a.c. mains voltage means that 240 V is the r.m.s. voltage of the mains.
The peak voltage E_p is 240 $\times$ $\sqrt{2}$ = 340 V. The 240 V a.c. mains thus produces
the same power as would be obtained from a steady voltage of this value.

26.5   THE TRANSFORMER

A transformer is illustrated diagrammatically in Fig. 26.4. The two coils shown
are wound on a laminated soft-iron core.

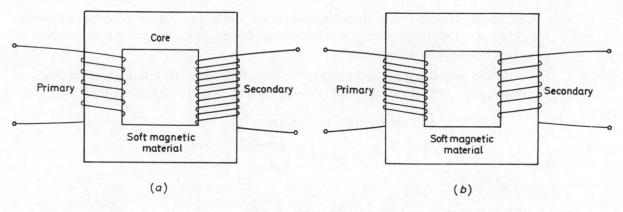

Fig. 26.4 Transformer

When an alternating current passes through the primary an alternating magnetic field is set up in the core. Since the flux through the secondary coil is changing, this induces an e.m.f. in it which is also alternating. The size of this induced e.m.f. will depend on the e.m.f. applied to the primary and on the relative numbers of turns in the two coils:

$$\frac{\text{secondary e.m.f.}}{\text{primary e.m.f.}} = \frac{\text{number of turns in secondary}}{\text{number of turns in primary}}$$

If the ratio of the number of turns is greater than one, the secondary e.m.f. will be greater than the primary and we have a *step-up* transformer. A ratio less than unity gives a *step-down* transformer.

A transformer is designed to minimise energy losses. The windings are composed of low-resistance copper coils to reduce the heating losses. The core is laminated, that is constructed of layers of magnetic material fixed together and insulated from each other by varnish or oxide coatings. This ensures that, although e.m.f.'s are induced in the core, as well as the secondary coil by the changing magnetic field, currents cannot flow between the laminated layers. Thus the total current in the transformer core is kept to as small a value as possible, as are the energy losses due to its heating effect.

Efficient core design also means that all the magnetic flux produced by the primary passes through the secondary. In practice this is best achieved by winding the primary and secondary on top of one another. In a good transformer energy losses are small and we may assume that:

secondary power output = primary power input.

26.6 POWER TRANSMISSION

One of the main advantages of alternating current is that it can easily be changed from one voltage to another by a transformer, with little loss of energy. For this reason, electric power is generally conveyed by alternating current, as it can be transformed to a very high voltage and transmitted over large distances, with small power losses as explained below. This has two advantages. Firstly, electricity can be generated where water power, coal and oil are easily obtainable or at conveniently sited nuclear power stations, and conveyed to all parts of the country by high voltage overhead power lines. Secondly, power is easily made available wherever the peak demand occurs. For example, during the day power which has been generated in an area that has little industry may be used in a more industrialised area that has at this time a greater demand.

In great Britain electricity is generated at 11 000 volts and then stepped up to as much as 400 000 volts by transformers. It is subsequently stepped down in stages at substations in the neighbourhood where the energy is to be consumed. The reason electrical energy is transmitted at such high voltages can clearly be seen from the following calculation.

Find the power wasted as heat in the cables when 10 kilowatts is transmitted through a cable of resistance 0.5 ohm at (a) 200 volts; (b) 200 000 volts.

(a) The current is given by the equation:

$$\text{current} = \frac{\text{power}}{\text{voltage}} = \frac{10\,000}{200}$$

$$= 50\,\text{A}.$$

Therefore the power lost in the cable $= I^2 R$
$$= 50^2 \times 0.5 \text{ watts}$$
$$= 1250 \text{ watts}.$$

(b) The current $= \dfrac{\text{power}}{\text{voltage}} = \dfrac{10\,000}{200\,000} = 0.05\,\text{A}.$

Therefore the power lost in the cable $= I^2 R$
$$= 0.05^2 \times 0.5 \text{ watts}$$
$$= 0.00125 \text{ watts}.$$

At 200 volts more than 10% of the energy transmitted is wasted in heating the cable. At 200 000 volts this energy loss is reduced by a factor of a million and is negligible.

27 Electron beams

27.1 THERMIONIC EMISSION

Metals contain many electrons which are loosely attached to their atoms. If a wire is heated to a high temperature the extra energy given to the electrons enables them to break away from the metal structure and exist outside as an electron cloud. This is called **thermionic emission.**

Strontium and barium are good electron emitters but are not suitable to be made into a thin wire. Tungsten, however, can be made into a very thin wire and can withstand high temperatures without melting. Thus tungsten is used as the base of the wire and its surface coated with barium or strontium which emit well at these temperatures. The wire is heated electrically, usually by using a 6.3 V supply which can be obtained from either a d.c. or an a.c. source.

27.2 THE DIODE

The electrons released from a wire by thermionic emission form a cloud around it which prevents further emission. This cloud may be removed by placing a plate near to the wire and connecting a steady voltage between them, so that the plate (anode) is more positive than the wire (cathode). It is necessary to place this whole arrangement in an evacuated tube so that the passage of the electrons is not inhibited by the presence of air molecules. This arrangement is known as a **diode**.

The characteristic action of a diode may be investigated using the circuit shown

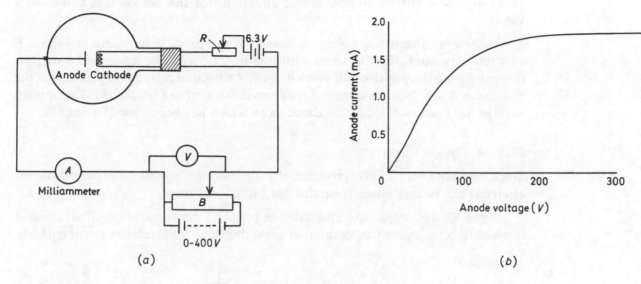

Fig. 27.1 Diode

in Fig. 27.1(*a*). The voltage applied to the diode is varied using the source *B*, and the resulting current is registered on the milliammeter. Fig. 27.1(*b*) shows a typical set of results.

The actual values of the voltage applied and the current obtained will depend on the precise diode used. However, the general shape of the graph will be similar in all cases. When no voltage is applied between the anode and cathode the electrons released by thermionic emission remain near the cathode. If a small positive potential difference is applied between the anode and cathode, some of the electrons in the cloud move across the empty space to the anode. Here they flow through the anode circuit back to the cathode, and a small current is registered.

The electrons crossing to the anode create a negative charge in the space between anode and cathode. This negative space charge repels some electrons back to the cathode. Thus not all the electrons emitted from the cathode reach the anode when the potential difference is small. However, as the potential difference increases, a larger proportion of the electrons emitted do reach the anode. At large voltages all the electrons emitted per second reach the anode and increasing the voltage further will not increase the current. The maximum current is known as the *saturation current*.

No current flows when the anode potential is negative with respect to the cathode. All the electrons emitted are repelled back to the cathode. The diode will

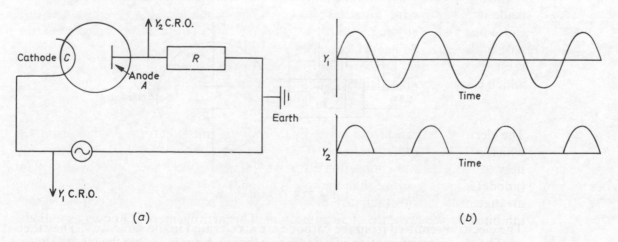

Fig. 27.2

thus only allow current to flow in one direction; for this reason it is known as a **valve**.

Suppose an alternating voltage is connected to a diode valve with a resistance R of a few thousand ohms in series with it (Fig. 27.2(a)). The diode will only pass current during the positive half of each cycle. An oscilloscope connected across the resistance R will show a voltage of the form illustrated in Fig. 27.2(b). The applied voltage has been rectified by the diode valve which has been used as a 'rectifier'.

27.3   CATHODE RAYS (ELECTRON BEAMS)

Since electrons are easily produced by thermionic emission, experiments on electrons can be conveniently carried out using this source.

In the Maltese cross tube illustrated in Fig. 27.3 the anode A is cylindrical and is maintained at a positive potential of a few thousand volts relative to the cathode

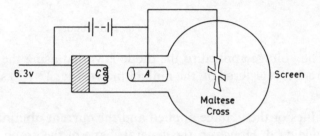

Fig. 27.3 Maltese cross tube

C. This part of the apparatus is called an *electron gun*. The electrons are accelerated in the space between the cathode and anode and pass through the cylindrical anode. The Maltese cross shown is at right angles to the beam of cathode rays and is connected to the anode. The cross is thus at the same potential as the anode and the electrons move between the two at a constant speed (that is, they are not accelerated). When the electrons strike the screen some of their kinetic energy is converted into light, thus showing the position of the beam. A sharp shadow of the cross is seen on the screen, suggesting that the electrons emitted from the cathode travel in straight lines along the tube.

27.4   THE EFFECT OF ELECTRIC AND MAGNETIC FIELDS

The deflection tube illustrated in Fig. 27.4 is similar to the Maltese cross tube, except that it contains two horizontal plates P and Q instead of the cross.

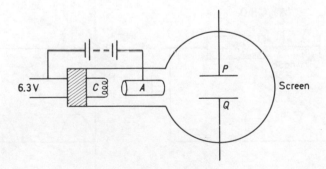

Fig. 27.4 Deflection tube

The electrons emitted from the cathode are accelerated in the same way. They then pass between P and Q before striking the centre of the screen. However, if a large potential difference is connected between P and Q, the electron beam will be deflected. If, for example, P is positive relative to Q the electrons will be attracted to it and the beam will be deflected towards it. The beam is deflected downwards if Q is the more positive plate. The amount of deflection depends on the potential difference.

If two bar magnets are placed, one each side of the tube, with opposite poles facing, the beam will be deflected. This is the same effect as that responsible for the force on a current-carrying wire placed in a magnetic field and thus for the rotation of an electric motor. In place of the bar magnets, the magnetic field may be produced by using two vertical coils, one each side of the tube. If a current of between one and two amps is passed through these coils connected in series, a magnetic field results in the space between them. It has already been seen, in unit 25.5, that a beam of charged particles passing through a magnetic field will experience a force at right angles to the field and their direction of travel. In the deflection tube the force acting on the electrons is initially either up or down, but, as the beam is deflected by this force, the force remains at right angles to the beam. The amount of deflection in a magnetic field depends on the strength of the field; in the case of the coils this depends on the current passing through them.

There is one further common type of deflection tube, known as the fine beam tube (Fig. 25.10). It contains a conical metal anode A, with a hole at the top over the cathode C. When a potential difference of a few thousand volts is connected between the two an electron beam is emitted vertically. The tube has a very small amount of hydrogen in it. The fast-moving electrons produce a fine beam of light as they ionise the hydrogen molecules. The beam shows the path of the electrons. The electrons may be deflected by connecting a potential difference between the two plates D_1 and D_2 just above the anode, or by passing an electric current through the two large coils placed one each side of the tube. In the latter case the beam of electrons may be deflected into a closed circle as the force is always at right angles to the direction of travel of the electrons.

27.5 ENERGY AND VELOCITY OF ELECTRONS

The glow of light in the deflecting and fine beam tubes is derived from the kinetic energy of the electrons leaving the anode. The electrons gain this kinetic energy by the conversion of the electrical potential energy between the cathode and anode. Using the principle of the conservation of energy the velocity of the electrons can be calculated. If an accelerating potential of 4000 volts is used the velocity of the electrons on leaving the anode is about 10^7 m/s or about 3% of the velocity of light.

27.6 CATHODE RAY OSCILLOSCOPE

The cathode ray tube shown in Fig. 27.5 is very similar to the deflection tube illustrated in Fig. 27.4. It relies on exactly the same principle for its operation but has an additional pair of plates X_1 and X_2 so that the electron beam may be deflected horizontally as well as vertically.

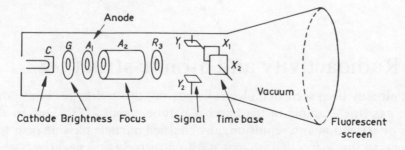

Fig. 27.5 Cathode ray oscilloscope

The cathode C emits electrons which are accelerated to the anode A by a high positive potential difference. In practice the anode usually consists of more than one plate or cylinder (A_1, A_2) so that it behaves like an electron lens and is able to focus the beam. The plate G is slightly negative compared to the anode. The number of electrons reaching the screen is controlled by how negative G is.

The cathode ray oscilloscope is excellent for use as a voltmeter. As no charge passes between the plates, when the beam is deflected, the tube draws no current from the component whose voltage it is recording; that is it has an extremely high resistance.

If a d.c. voltage is to be measured it is connected between the plates Y_1 and Y_2. The spot on the screen will move a certain distance vertically which should be measured. If the oscilloscope has a calibrated scale this reading may immediately be converted to volts. If it is not calibrated a battery of known voltage should be connected to the oscilloscope and the deflection it causes recorded.

If it is required to measure an a.c. voltage, this should be connected to the plates Y_1 and Y_2. The length of the vertical line which results gives the value of the peak to peak voltage. However, if it is desired to see how the value of the voltage changes with time, the time base facility should be used. This provides a changing voltage connected to the X-plates, thus deflecting the beam horizontally. The spot thus repeatedly moves across the screen at a speed dictated by a switch controlling the time base frequency. Thus, as the spot moves up and down in response to the a.c. voltage connected to the Y-plates, it moves horizontally at a steady speed. A graph (Fig. 27.2 (b) Y_1) is obtained with voltage as the Y-axis and time as the X-axis.

The cathode ray oscilloscope may be used to display any a.c. voltage; for example that across a resistor or that given by a microphone when sound falls on its diaphragm. It may also be used to examine the small voltage associated with the human heartbeat. The time base mechanism makes the oscilloscope suitable for use as a timing device. A pulse can be displayed on the screen at the instant it is emitted by a radar system and again when it is received back after reflection by an object in the earth's atmosphere or in the space beyond. The distance between the two pulses on the screen gives the time taken for the return journey of the pulse. As the velocity of radio waves is known to be 3×10^8 m/s, the distance between the emitter and the reflecting object can be calculated.

The cathode ray tube is the basis of a television set. The time base causes the spot to cross the screen many times in quick succession, each time a little below the last, so that the whole screen is covered. In Great Britain and many other countries 625 lines are drawn on the screen in 1/25 second in this way. As the spot moves across the screen its intensity is varied by the incoming signal, thus causing a picture to be 'painted' on the screen. The whole process is then repeated. Thus 25 slightly different pictures appear on the screen every second, giving the impression of continuous movement.

28 Radioactivity and atomic structure

It has already been seen that electrons passing through a gas ionise the molecules of the gas (for example in the fine beam tube – unit 27.3). Due to the electrical forces of attraction and repulsion, any charged particle passing near to a molecule will tend to ionise it – that is, split it into positively and negatively charged parts. This property of charged particles may be used to detect their presence.

28.1 RADIATION DETECTORS

One of the most widely used detectors is the Geiger-Müller (GM) tube shown in Fig. 28.1.

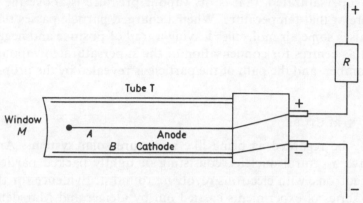

Fig. 28.1 Geiger tube

It consists of a small closed glass tube *T*, with a thin mica end window *M*, and contains a gas such as argon at about half atmospheric pressure. A thin wire *A*, which acts as the anode, passes down the centre of the tube and is insulated from it. The inside of the tube is coated with a conductor and forms the second electrode *B*. A potential difference of about 450 volts is applied between the electrodes.

When a charged particle enters the tube through the thin mica window some argon atoms are ionised. Many more atoms throughout the tube then become ionised. The negative ions produced are attracted towards the central wire *A*, the positive ions going towards *B*. A current is thus obtained in the circuit for a short time. It is called a pulse of current and causes a voltage pulse in the high resistance *R*. The 450-volt power supply and the resistance *R* are contained in an electronic unit such as a scaler, which counts individual pulses, or a ratemeter, which gives the average count rate.

The presence of X-rays and γ-rays may also be detected using a GM tube.

A cloud chamber may also be used to detect the presence of charged particles. One type – the diffusion cloud chamber – is shown in Fig. 28.2.

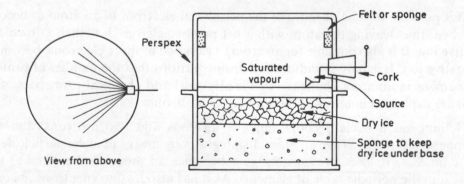

Fig. 28.2 Diffusion cloud chamber

It consists of a cylindrical perspex chamber with a detachable lid. The base of the chamber may be cooled by placing dry ice in a container below it. The sponge just below the lid of the chamber is first moistened with methyl alcohol. The base of the container is then removed and dry ice packed under the base of the chamber. The sponge is replaced to keep the dry ice in contact with the base of the chamber. It is important that the chamber is levelled, thus reducing convection currents.

The dry ice cools the base of the chamber to about $-60°C$. Alcohol evaporates continuously from the sponge below the lid, which is at room temperature, and the vapour sinks continuously to the bottom of the chamber. Just above the cold base the vapour is supersaturated, that is, its vapour pressure is above the saturated vapour pressure at that temperature. When a charged particle passes through the chamber it ionises some air molecules leaving a trail of positive and negative ions. These ions act as centres for condensation of the supersaturated vapour near the base of the chamber, and the path of the particle is revealed by the droplets which form.

28.2   ATOMIC STRUCTURE

Atoms may be regarded as being like miniature solar systems. An atom is thought to have a central nucleus, consisting of tightly packed particles called protons and neutrons, with electrons revolving round it. Evidence for this model comes from a series of experiments carried out by Geiger and Marsden at Manchester University in 1911. Alpha particles emitted from a radioactive source were scattered by a thin foil of gold. A very small fraction of the particles were scattered through very large angles (some through more than 90°) and Rutherford suggested that these had come very close to a high concentration of positive charge (an atomic nucleus) and had been deflected by the large electrical force of repulsion.

In the nucleus of an atom each proton has a positive charge equal in magnitude to the charge on an electron, and the number of protons in the nucleus of an atom is equal to the number of electrons in the atom. The atom is therefore electrically neutral. The chemical properties of an atom and hence the element to which it belongs are dictated by the number and arrangement of electrons in it, and this in turn depends on the number of protons in the nucleus, known as the **atomic number**, denoted by Z.

Neutrons are similar in mass to protons, both being nearly 2000 times as massive as electrons. The mass of an atom is thus almost entirely due to the masses of the protons and neutrons contained in its nucleus. The total number of protons and neutrons in the nucleus of an atom is known as the **atom's mass number**, denoted by A. Thus if the number of neutrons is denoted by N we have

$$A = Z + N.$$

It is possible for one or more of the peripheral electrons of an atom to become detached, thus leaving the atom with a net positive charge. It is then known as a positive ion. It is also possible for an atom to gain one or more electrons, becoming a negative ion. It is only in radioactive disintegrations that the number of protons or neutrons in an atom changes. All the electrical and chemical properties of an atom are explained in terms of the transfer of electrons.

Helium has a nucleus containing two protons and two neutrons, and two peripheral electrons. When an atom disintegrates by means of **alpha particle decay** (see unit 28.4) it loses two protons and becomes an atom of the element two below it in the periodic table of elements. As it has also lost two neutrons, its mass number has fallen by four.

Beta decay seems to take place by a neutron changing to a proton, which remains in the nucleus, and an electron which is emitted. The atom concerned has gained a proton and thus becomes an atom of the element one up in the periodic table. Its mass number has not changed.

28.3   ISOTOPES

From the previous discussion it can be seen that two atoms of an element may exist (that is the atoms have the same number of protons) which have different numbers of neutrons and hence different mass numbers. These atoms are said to be **isotopes of the same element**.

At the lower end of the periodic table the number of neutrons present in an atom is equal to, or nearly equal to, the number of protons. Elements of high mass number consist of atoms with very many more neutrons than protons. It is these atoms which are more likely to disintegrate. It seems that a large number of excess neutrons in an atom leads to instability and the likelihood of spontaneous disintegration. Generally speaking it is the elements of high mass number which exhibit this property of radioactivity.

28.4 RADIOACTIVITY

A number of naturally occurring substances emit particles or radiations which ionise gases. Marie Curie and her husband did much of the early work on radioactive substances and showed that amongst the most active were those containing uranium, polonium and radium.

By 1899 Rutherford had shown that the particles or radiations emitted by radioactive substances fell into three categories which he called **alpha (α)**, **beta (β)** and **gamma (γ) rays**.

Alpha rays are helium ions – that is, helium atoms which have lost two electrons, and hence have a positive charge. From a particular radioactive substance they are all ejected with approximately the same velocity and hence kinetic energy. They have a range of a few centimetres in air at atmospheric pressure, but most are stopped by a thick sheet of paper. Like all charged particles they lose their energy by continuous ionisation and the fact that they produce many ions per centimetre of path means that they travel relatively short distances. Being charged particles they are deflected by both electric and magnetic fields.

Beta rays are streams of high energy electrons similar to cathode rays but travelling much faster. They are emitted with velocities approaching that of light (3×10^8 m/s) and, as they do not form such a high density of ions along their track, their range is greater than that of alpha rays. Beta rays are negatively charged and are thus deflected in the opposite direction to alpha particles in both electric and magnetic fields.

Gamma rays are very short wavelength radiations similar to X-rays, which can penetrate several centimetres of dense metal such as lead. They do not produce continuous ionisation but lose their energy in one interaction with a molecule. The fact that they cannot be deflected by electric or magnetic fields indicates that they are not charged particles.

An idea of the range of alpha, beta and gamma rays may be obtained by placing sources emitting these separate rays in front of a GM tube connected to a scaler. The number of counts recorded by the scaler in a given time is noted with different absorbers (sheets of paper, aluminium or lead of different thicknesses) between the source and the tube. Alphas are stopped by a thin sheet of paper. It is found that gamma rays in particular and beta rays to a lesser extent are very difficult to stop or screen, and thus may present a safety hazard. Gamma ray sources are kept in lead containers the walls of which are often several centimetres thick.

If the GM tube and scaler are switched on in the absence of a source and the absorbers, some counts will be recorded. This count rate, which is slow, is due to radiation which is always present in the atmosphere. It is called background radiation, and should be subtracted from the counts recorded, when a source is present, to give the true count rate due to the source.

28.5   RADIOACTIVE DECAY

By 1903 Rutherford had come to the conclusion that radioactivity is the result of the spontaneous distintegration of an atom during which it emits an alpha or beta ray. Simultaneously the atom changes into one of another element which may itself be radioactive. The following are examples of such disintegrations.

$$^{238}_{92}U \rightarrow {}^{4}_{2}He + {}^{234}_{90}Th \qquad \alpha\text{-decay}$$

$$^{234}_{90}Th \rightarrow {}^{0}_{-1}e + {}^{234}_{91}Pa \qquad \beta\text{-decay}$$

$$^{234}_{91}Pa \rightarrow {}^{0}_{-1}e + {}^{234}_{92}U \qquad \beta\text{-decay}$$

The numbers above the symbols are the mass numbers of the atoms; those below are the atomic numbers.

It is impossible to predict which particular atom in a sample will change in this way; that is, the disintegration of an atom is a random event. However, in a sample

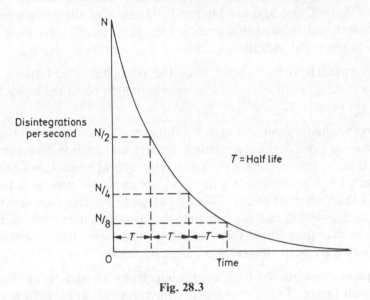

Fig. 28.3

containing a large number of atoms, the number which disintegrate each second will depend on the number of undecayed atoms left. Thus if the number of disintegrations per second is recorded (by counting the number of particles emitted) and plotted against time an exponential graph results (Fig. 28.3).

As can be seen from this graph, an infinite time is taken for all the atoms of a sample to disintegrate. It is therefore meaningless to talk about the 'life' of a sample. However, **the time taken for half the undecayed atoms in any given sample of the substance to disintegrate is finite, and the same whatever the number of undecayed atoms present initially.** This time T is known as the **half-life** of the substance. It is different for different radioactive materials.

28.6   SAFETY

Exposure of the body to radiation from atomic disintegrations can have undesirable effects of a long- or short-term nature. The precise result of exposure depends on the nature of the radiation, the part of the body irradiated and the dose received. The hazard from alpha particles is slight, unless the source enters the body, since they cannot penetrate the outer layers of skin. Beta particles are more penetrating, although most of their energy is absorbed by surface tissues. Gamma rays present the main external radiation hazard since they penetrate deeply into the body.

Radiation can cause immediate damage to cells, and is accompanied by radiation burns (redness of skin followed by blistering and sores, the severity of these depending on the dose received), radiation sickness and possibly death. Effects such as cancer and leukemia may appear many years later, due to the uncontrolled multiplication of some cells set off by exposure to radiation. Hereditary effects may also occur in succeeding generations due to genetic changes. The most susceptible parts of the body are the reproductive organs and blood-forming organs such as the liver.

Because of the hazards it is essential that the correct procedure be adopted when using radioactive substances.

Briefly it is as follows:
1. Sources should only be held with forceps, never with the hand. This avoids any possibility of some of the substance transferring to the surface of the skin or lodging under a nail.
2. Any cuts on the hand should be covered before using sources.
3. Sources should not be pointed towards the human body.
4. Sources should be returned to their container as soon as they are no longer needed for an experiment. If possible, permanent storage should be within two containers.
5. A check should be made at the end of an experiment to see that all sources are present in their allotted containers.
6. Mouth pipettes should never be used with liquid sources.
7. Do not remain unnecessarily in the region of a radioactive source.

Section III
Hints for candidates taking Physics examinations

All questions must be attempted in multiple choice, or as they are sometimes called 'fixed response', examination papers. It is therefore unnecessary and a waste of time to read the paper through before starting to write down your answers. However, as you come to each question do read it carefully before answering it.

Start at Question 1 and work steadily through the paper doing all the questions you can answer fairly easily. Do not delay, at this stage, over questions which you find difficult. Working in this way you should be able to reach the end of the paper in a little over half the time allowed for it, having answered about two-thirds of the questions. Now return to the beginning of the paper and tackle the questions you found difficult the first time through, but still leave out questions about which you have little or no idea. You should aim to reach the end of the paper five or ten minutes before the end of the examination.

Now is the time to return to the few questions still unanswered and make an *intelligent* guess in each case. Generally speaking you will be able to eliminate two or three of the alternative answers to each question, but you may not be sure which one of those left is correct. If you guess, at this stage, you will have a one in two, or one in three, chance of being right. However *wild guessing is useless* and you should not guess at any answers until the last few minutes of the examination. *Do not leave any questions unanswered.* Examination Boards do not deduct marks for wrong answers.

If you have any time left, when you have answered all the questions, check through your answer sheet, bearing in mind the following points:

1 All questions should have an answer.
2 No question should have more than one answer – this can occur by mistake when filling in the sheet.
3 Any alteration should be such that the incorrect answer has been completely rubbed out.
4 A soft pencil should always be used for completing the answer sheet.

Questions on multiple choice examination papers fall into three basic types. Questions of the first type can be found under the heading **Type 1** in the self-test units (pages 134–136). The questions occur in groups with the same alternative answers for each question within the group, and the candidate has to select the most appropriate answer for each question. Within each group of questions each answer may be used once, more than once, or not at all.

In **Type 2** each question has its own group of alternative answers and the candidate must select the best answer to each one (see pages 136–141). Examples of the third type of question, known as multiple completion, are given under the heading **Type 3** (see pages 141–143). Each question is followed by three statements, one or more of which may be correct. The candidate has to decide which statement(s) are correct and select the letter which represents the combination of correct statements. When tackling these questions it is worth placing a mark against each correct statement, as one goes along, for use when choosing the correct letter at the end of the question.

Free response questions can be defined as any question where the candidate has to make a written answer rather than just choose from the answers given. There is a wide range of free response questions ranging from short answers (where the answer may be a few words or sentences) through to longer structured questions.

A selection of short questions is given in Section V (see pages 144–160). In this type of question space for the answers is often left on the question paper. If this is the case, the space left is a guide, but only a guide, to the length of the answer required. Do not feel that you *have* to fill the space, but if you find your answer is much too short or much too long think again. Whether space is left to answer on the question paper or not, your answers should contain all the relevant points stated in a logical order, and concisely expressed.

LONGER QUESTIONS

A selection of longer questions is given in Section V (see pages 160–216). Examination papers containing this type of question vary considerably in their style and length of question, as well as how many questions they require the candidate to answer. It is essential to read the instructions printed at the beginning of the paper to find out which questions, if any, are compulsory, and how many of the others should be answered. Do not answer more questions than instructed.

If you are presented with a choice of question or parts of a question, it is essential that you read through *all* the alternatives before making your choice. It is not only important to make the correct choice for you, but then to do *first* that question or part question which you consider easiest. You are more likely to do well on the more difficult questions if you have already successfully completed some easier ones. However be sure to number clearly the answers to questions and parts of questions.

It is also important that you divide up your time for the examination correctly. If, for example, four questions must be answered in two hours, then you cannot afford to spend more than 35 minutes on any one question. In fact the first question or two should each take you slightly less than half an hour rather than more.

Generally more time is allowed for answering each question than is required simply to write the answer down. It is therefore sensible to spend a few minutes planning your answer. First read the question carefully, then think which physical principles are involved. If all or part of the question requires a descriptive answer, it is worth jotting down in rough the points to be included in your answer, and deciding on the order in which they should be presented, before starting to write the answer. Your answer should be written down accurately and concisely. You will lose marks by making vague statements, and you will run short of time if you use an *excessive* number of words. Make sure the words you use are essential and have meaning. If you have to describe an experiment, not only describe the apparatus but how any measurements are made. For example, if a stopwatch is used to time 20 swings of a pendulum, say just that. By all means use diagrams to illustrate your answer, but they should be drawn clearly and neatly and labelled; otherwise they are useless.

If a calculation is involved, make sure you write down the physical principle you are using, in words or in the form of an equation, before starting on the arithmetic. At the end of the calculation see that your answer is physically reasonable. For example, it would not be reasonable to give the mass of a man as 600 kg; it is more than likely that you have made an arithmetical mistake and the the answer should be 60 kg. If in doubt about your answer, check your Physics and your Arithmetic. Make sure your answer is followed by the correct unit.

Most Examination Boards now print the mark allocation at the right of the question. If a question has several parts, the marks are a guide to the relative importance of the parts, and thus the time you should spend on each.

Section IV
Self-test units

This section contains a selection of multiple choice questions and is designed to test your knowledge of the material used in the book. Before attempting these questions you must read Section III (pages 132–133).

(Answers on page 143.)

Type 1

Each question below has five possible answers labelled A, B, C, D, E. For each numbered question select the one most appropriate answer. Within the group of questions each answer may be used once, more than once or not at all.

Questions 1–3

    A newton per second
    B newton metre
    C newton metre per second
    D newton second
    E kilogram metre per second^2

    Which one of the above is a unit in which each of the following physical quantities might be measured?

1 momentum
2 work
3 power

Questions 4–7

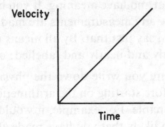

The graph shows how the velocity of a moving object starting from rest changes with time

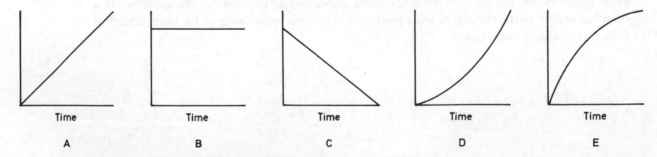

A	B	C	D	E

    Select from the graphs A–E above the one which most nearly represents the shape of each of the following graphs for the moving object.

4 the force acting on the body against time
5 distance travelled against time
6 velocity against time
7 kinetic energy against time

Questions 8–12

The following are five properties of a body made of the same metal throughout.

A mass B volume C weight D relative density E surface area
Which of the properties

8 is a vector quantity?
9 would change if the body were taken to the moon, its temperature remaining constant?
10 has no units?
11 would not change if the body were cut in half and only one half were considered?
12 needs to be known when calculating the upthrust when the body is immersed in a liquid of known density?

Questions 13–15
The following are five units –

A kilogram   D newton
B kilogram metre per second   E metre per second^2
C joule

Which of the above units are suitable for measuring the following quantities?

13 The kinetic energy of a cricket ball
14 The momentum of a cricket ball
15 The weight of a cricket ball

Questions 16–18
Which of the following words best describes the particles in question?
A protons   B molecules   C ions   D electrons   E neutrons

16 Carbon dioxide gas at room temperature
17 Particles emitted from a hot filament in the tube of a cathode ray oscilloscope
18 Alpha particles emitted by a radioactive substance

Questions 19–21
Five types of radiation are listed below.

A ultraviolet radiation   D β-radiation
B infra red radiation   E ultrasonic radiation
C γ-radiation

Select the type of radiation to which each of the following applies.

19 It has a wavelength slightly shorter than that of visible light and it can cause certain materials to fluoresce.
20 It has an extremely short wavelength, much shorter than that of visible light, and can be emitted from the nucleus of an atom.
21 It has a wavelength slightly longer than that of visible light and may be used for heating.

Questions 22–24
Resistors of 1 ohm, 1 ohm and 2 ohms are connected in arrangements lettered A to E below, and a potential difference is applied between *X* and *Y*.

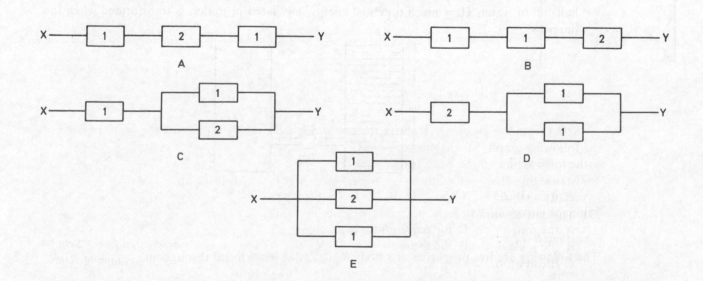

22 Which arrangement has the smallest total resistance?

23 Which arrangement has a total resistance of 2.5 ohms between *X* and *Y*?

24 Which arrangement would have the largest current through the 2 ohm resistor?

Type 2

Each of the following questions or incomplete statements is followed by five suggested answers labelled A, B, C, D, E. Select the best answer in each case.

25 Which of the graphs best represents the velocity-time graph of a ball thrown vertically upwards to a considerable height and which then returns to the thrower.

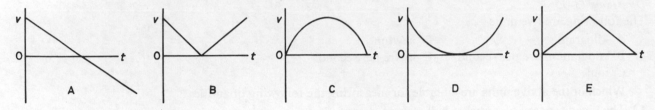

26 If the force acting on a moving object is constant in magnitude and is always perpendicular to the direction of motion of the object, then the object

  A accelerates with uniform acceleration.

  B decelerates with uniform deceleration.

  C moves in an elliptical path.

  D continues in a straight line.

  E moves in a circular path.

27 A 5 kg mass is travelling with a speed of 5 m/s. It is brought to rest in 0.5 s. The average force measured in newtons acting on it to bring it to rest is

  A 50   B 25   C 10   D 2.5   E 0.5

28 A car of mass 800 kg which is moving east at 50 m/s collides head on with a van of mass 1600 kg which is moving west at 10 m/s. If the vehicles stick together, the wreck will

  A be stationary.   D move east at 10 m/s.

  B move east at 20 m/s.   E move west at 10 m/s.

  C move west at 20 m/s.

29 8 kilograms of water is tipped from a bucket at the top of a building of height 200 m and forms a static pool on the ground. What will be its change in energy?

  A 1 600 000 J gain   D 16 000 J loss

  B 16 000 J gain   E 1 600 000 J loss

  C no change

30 The tank in the diagram is filled to a depth of 1 m with 10 kg of water. The tank is joined to an identical tank by a pipe with a tap in it. When the tap is opened each of the tanks becomes half-full of water. How much potential energy, measured in joules, is transformed when this happens?

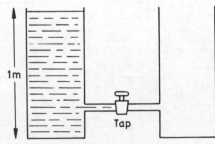

  A 0   B 2.5   C 5   D 10   E 25

31 The unit of work is

  A the watt.   D the newton/metre.

  B the joule.   E the metre.

  C the newton.

(Associated Lancs Schools Examining Board)

32 A ship is sailing slowly due East, and a passenger is on the deck at *P*. The points *A*, *B*, *C*, *D* and *E* are marked on the deck. Towards which of these points should the passenger walk at a suitable speed in order to move due North?

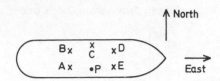

33 The diagram below shows a uniform beam *XY* of length 1 m supported at *P* where *PX* = 20 cm. The beam weighs 5 N.

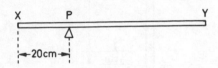

What downward force exerted at *X* will make the beam balance?

A 5 N   B 7.5 N   C 12.5 N   D 20 N   E 25 N

34 The centre of mass of a triangular shaped piece of thin cardboard lies on a line

A from one vertex perpendicular to the opposite side.
B through the mid-point of one side and perpendicular to that side.
C through the mid-point of one side and through the opposite vertex.
D through the mid-point of one side and parallel to an adjacent side.
E bisecting the angle at a vertex.

35 A single string pulley system is illustrated below. A load of 10 newtons can be raised by an effort of 4 newtons.

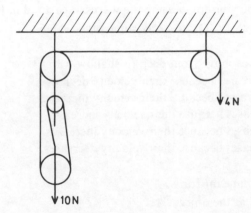

If the system is frictionless, the weight of the lower pulley block is –

A 2 N   B 2.5 N   C 3 N   D 6 N   E 12 N

36 A boy of mass 40 kg balances evenly on two stilts, each having an area of 8 cm^2 in contact with the ground. The pressure exerted by one stilt is

A 50 N/cm^2 B 40 N/cm^2 C 25 N/cm^2 D 5 N/cm^2 E 2.5 N/cm^2

37 The mass of an aluminium cube, in air, is 20 g. When suspended in oil its apparent mass is 15 g, and when suspended in water its apparent mass is 13 g.
The relative density of oil is

A $\frac{3}{4}$ B $\frac{15}{13}$ C $\frac{13}{15}$ D $\frac{7}{5}$ E $\frac{5}{7}$

38 An aluminium cube has an apparent mass of 24 grams in air, 16 grams when fully immersed in water and 17 grams when fully immersed in methylated spirits. The relative density of the methylated spirits is

A $\frac{17}{24}$ B $\frac{7}{8}$ C $\frac{16}{17}$ D $\frac{17}{16}$ E $\frac{8}{7}$

39 A bar of copper is heated from 290 K to 300 K. Which of the following statements is NOT true?

A Its length will increase slightly.
B Its electrical resistance will increase slightly.
C Its density will increase slightly.

D Its mass will remain unchanged.

E Its weight will remain unchanged.

40 In an uncalibrated thermometer the length of mercury in the stem is 4 cm when the thermometer is in pure melting ice, and 44 cm when in pure steam, both under standard pressure.

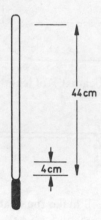

When placed in a liquid of unknown temperature, the length is 48 cm. The temperature of the liquid, measured in °C, is therefore

A 48   B $109\frac{1}{11}$ C 110   D 140   E 160

41 An electric heating wire is immersed in 0.05 kg of oil in a calorimeter of negligible heat capacity. The temperature of the oil rises from 20°C to 50°C in 100 seconds. If the specific heat capacity of oil is 2000 joules per kilogram degree C, the power supplied by the heating coil in watts, is

A 20   B 30   C 50   D 3000   E 300 000

42 In a ripple tank, waves travel a distance of 45 cm in 3s. If the distance apart of the crests is 3 cm, the frequency of the vibrator causing the waves, measured in Hz, is

A 5   B 7.5   C 11.25   D 20   E some other value.

43 What happens when waves move from deep to shallow water?

A Their wavelength decreases because their velocity decreases.

B Their wavelength decreases because their velocity increases.

C Their frequency increases because their velocity increases.

D Their frequency decreases because their velocity increases.

E Their frequency decreases because their velocity decreases.

44 The image formed by a plane mirror is

A real and the same size as the object.

B real and nearly the same size as the object.

C virtual and the same size as the object.

D virtual and nearly the same size as the object.

E virtual and half the size of the object.

(Associated Lancs Schools
Examining Board)

45 Which one of the following must be moved in order to give the correct sequence of wavelengths?

A Radio waves

B Ultra-violet rays

C Infra-red rays

D Yellow light

E Gamma rays

46 Sound waves are different from light waves because

A sound waves need no medium.

B light waves need a medium.

C sound waves need a medium.

D light waves travel through glass.

E sound waves are not refracted.

(Associated Lancs Schools
Examining Board)

47 Which of the following statements about waves is true?

A Radio waves, light waves and sound waves will all travel through a vacuum.

B Radio waves and light waves will travel through a vacuum, sound waves will not.

C Sound waves will travel through a vacuum, radio waves and light waves will not.

D Light waves and sound waves will travel through a vacuum, radio waves will not.

E None of the waves will travel through a vacuum.

48 A loudspeaker gives out a note of frequency 100Hz (cycles per second). If the speed of sound is 330 m/s, then the wavelength of the sound, in metres, is

A 1.1   B 3.3   C 11   D 33   E 110

49 A rod of insulating material is given a positive charge by rubbing it with a piece of fabric, and the fabric is then tested for electric charge. You would expect the fabric to have

A a positive charge equal to that on the rod.

B a negative charge equal to that on the rod.

C a positive charge less than that on the rod.

D a negative charge greater than that on the rod.

E no charge.

Questions 50–51

Two coils of wire, of resistance 2 ohms and 3 ohms respectively are connected in series with a 10 volt battery of negligible internal resistance as shown

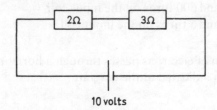

10 volts

50 The current through the 2 ohm coil, in amperes, is

A 0.5   B 2   C 5   D 20   E 50

51 The drop in potential across the 3 ohm coil, in volts, is

A 2   B 4   C 5   D 6   E 10

52 Which of the following would enable a milliammeter to be used as a direct current voltmeter?

A Connecting a large resistor in series.

B Connecting a small resistor in series.

C Connecting a large resistor in parallel.

D Connecting a small resistor in parallel.

E A milliammeter can never be adapted for measuring voltages.

53 A 24 watt 12 volt headlamp is lit by connecting it to a 12 volt battery of negligible resistance. The current flowing, in amperes, is

A 288   B 24   C 12   D 2   E 0.5

54 An electric fire is rated 250 V 1000 W. If electricity costs 2p per unit what is the cost of running the fire for 5 hours?

A 1p   B 2p   C 5p   D 10p   E 20p

(Associated Lancs Schools Examining Board)

55 The outer casing of an electric iron is generally connected to earth in order to

A prevent a serious electric shock.

B complete the circuit.

C prevent the fuse from burning out.

D protect the iron.

E allow the current to get away.

56 An ammeter is required to measure alternating current to a maximum of 10 amperes. For this purpose a moving coil instrument:

A should have a small resistance connected in series.
B should have a small resistance connected in parallel.
C should have a large resistance connected in series.
D should have a large resistance connected in parallel.
E would not be suitable.

57 A 12 V 36 W lamp is connected across the output of a transformer. The output has 60 turns

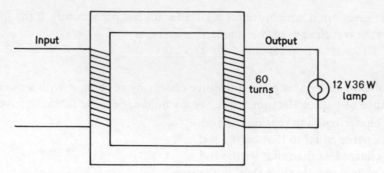

round the transformer. Which one of the following inputs to the transformer would enable the lamp to glow with normal brightness?

A a voltage of 12 V d.c. and 36 turns on the input coil.
B a voltage of 120 V d.c. and 600 turns on the input coil.
C a voltage of 12 V a.c. and 36 turns on the input coil.
D a voltage of 120 V a.c. and 600 turns on the input coil.
E a voltage of 120 V a.c. and 6 turns on the input coil.

Questions 58–59
In a school experiment a stream of electrons passes through a horizontal slit and strikes an inclined screen so that a trace is seen as indicated in the diagram.

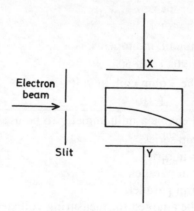

58 Which one of the following is the best explanation of the parabolic path?
A The electrons are falling under the influence of gravity.
B There is a magnetic field acting downwards between the plates.
C The electrons are slowing down and losing energy.
D Plate X has a positive potential relative to plate Y.
E Plate X has a negative potential relative to plate Y.

59 Which one of the following procedures is the most likely to change the trace to a horizontal line?
A increase the velocity of the electrons.
B apply a magnetic field in the direction of the beam.
C apply a magnetic field in the direction XY.
D apply a magnetic field in the direction YX.
E apply a magnetic field at right angles to both XY and the electron stream.

60 Atoms of atomic number (proton number) 92 and mass number (nucleon number) 234 decay to form new atoms of atomic number 90 and mass number 230. The accompanying emissions will be

A electrons.   B newtons.   C gamma rays.   D beta particles.   E alpha particles.

61 Thorium 232 with atomic number 90 decays to element X by ejecting an alpha-particle. Element X emits a beta-particle to become element Y. Element Y emits another beta-particle to become element Z. The mass number and the atomic number of element Z are

	Mass number	Atomic number
A	230	90
B	228	88
C	228	90
D	230	88
E	226	90

Type 3

For each of the questions, one or more than one of the responses 1, 2, 3, is/are correct. Select the one letter A, B, C, D, E, which represents the correct responses.

A 1, 2 and 3 are correct.
B 1 and 2 are correct.
C 2 and 3 are correct.
D 1 only is correct.
E 3 only is correct.

62 The acceleration due to gravity on the moon is less than that on the earth. Which of the following would have different values if measured on the moon and on the earth?

1. The time taken for a stone to fall through a given height.
2. The energy of a body moving at a given speed.
3. The momentum of a body moving at a given velocity.

63 To undo a tight nut a mechanic uses a spanner. This

1. reduces the force due to friction between the nut and bolt.
2. reduces the velocity ratio of the nut on its screw thread.
3. increases the moment of the force the mechanic can apply to the nut.

64 A crowbar is used to raise a load of 80 N, by applying a force of 20 N. The total length of the crowbar is 1 m and it is pivoted 20 cm from the load.

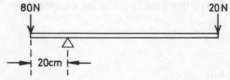

If it is regarded as a machine, then

1. The velocity ratio is 5.
2. The mechanical advantage is 4.
3. No energy is wasted in raising the load.

65 It is more painful to carry a heavy parcel if it is held by a string than if it is held by a thick strap because compared to the strap

1. the string exerts a greater force on the fingers.
2. the string has a greater tension.
3. the string exerts a greater pressure on the fingers.

66 The sketch shows a water manometer made from wide glass tubing and designed to compare the pressure of two gases P and Q.

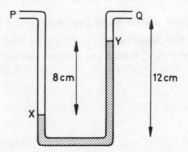

1. The pressure of gas P is greater than that of Q.
2. The maximum pressure difference that this manometer can measure is equal to that produced by a column of water of height 12 cm.
3. If the pressures at P and Q became equal then the surface at Y would move down 8 cm.

67 Water and water vapour are both at room temperature. The molecules of the vapour differ from those of the liquid in that the vapour molecules
1. are less dense.
2. move with a greater average speed.
3. have a larger average distance between them.

68 At the end of a hot day the wind tends to blow from the sea to the land. Which of the following contribute to this?
1. The specific heat capacity of water is greater than that of land.
2. Less radiation per second is absorbed by unit area of sea than land.
3. Convection currents occur in the air due to inequality of temperature over land and sea.

69 Pressure cookers cook food more quickly than do ordinary saucepans covered by a lid because
1. the increased pressure lowers the boiling point so that the water boils more easily.
2. the pressure forces heat into the food.
3. the temperature attained by the contents is greater than 100°C.

70 Water can be made to boil at a temperature of 101°C if
1. it is heated very quickly.
2. salt is added.
3. the pressure is increased.

71 In a thermos flask for keeping liquids hot, the vacuum reduces heat losses due to
1. conduction.
2. convection.
3. radiation.

72 An eclipse of the sun can take place when
1. there is a new moon.
2. the earth is in the shadow of the moon.
3. the earth is exactly between the moon and the sun.

73 The image of a distant scene produced on the film of a camera is
1. real.
2. diminished.
3. erect.

74 Both a lens camera and the human eye
1. contain a converging lens.
2. form a real image.
3. have a device to adjust the admission of light.

75 An electric lamp is labelled 12 volt 3 ampere. When it is operating correctly
1. the power of the lamp is 4 watts.
2. the resistance of the lamp is 36 ohms.
3. the energy consumed in 10 seconds is 360 joules.

76 A neutral atom of cadmium contains 48 protons and 68 neutrons.
1. Its mass number (nuclear number) is 68.
2. It has 48 electrons.
3. It is an isotope of an atom which has 48 protons and 66 neutrons.

For each of the following questions, one or more than one of the three statements (i)–(iii) is correct. You may find it helpful to put a tick against any statements which you consider to be correct, and then to select that one of the letters A–E which represents the number(s) you have ticked.

A (i), (ii) and (iii)
B (i) and (ii) only
C (i) and (iii) only
D (iii) only
E some other response

77 Brownian motion is observed in smoke. Which of the following statements about the motion is/are true?

    (i) It would be more vigorous if lighter smoke particles were used.

    (ii) It would be less vigorous if the temperature were lowered.

    (iii) It is due to smoke particles colliding with one another.

78 A coil of wire is rotated in a magnetic field. The axis of rotation is at right angles to the field

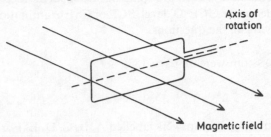

as shown. The e.m.f. induced will be smaller if

    (i) the coil is rotated more slowly.

    (ii) the coil has fewer turns.

    (iii) the magnetic field is decreased.

79 A beam of charged particles is moving in a circular path in a uniform magnetic field. Assuming that in each of the following cases each of the other factors remains unchanged, the radius of the circular path will be increased if

    (i) the strength of the magnetic field is increased.

    (ii) the velocity of the particles is increased.

    (iii) the mass of the particles is decreased.

80 A signal is applied to an oscilloscope and the trace appears as in Fig. (*a*). Which of the

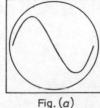

Fig. (*a*)

Fig. (*b*)

following is/are possible adjustment(s) to cause the trace in Fig. (*a*) to change to that in Fig. (*b*)?

    (i) The sweep speed of the time-base is doubled.

    (ii) The magnitude of the signal voltage is doubled.

    (iii) The frequency of the signal voltage is doubled.

Answers to self-test units

1	D	15	D	29	D	43	A	57	D	71	B
2	B	16	B	30	E	44	C	58	E	72	B
3	C	17	D	31	B	45	B	59	E	73	B
4	B	18	C	32	B	46	C	60	E	74	A
5	D	19	A	33	B	47	B	61	C	75	E
6	A	20	C	34	C	48	B	62	D	76	C
7	D	21	B	35	A	49	B	63	E	77	B
8	C	22	E	36	C	50	B	64	C	78	A
9	C	23	D	37	E	51	D	65	E	79	E
10	D	24	E	38	B	52	A	66	B	80	D
11	D	25	A	39	C	53	D	67	E		
12	B	26	E	40	C	54	D	68	A		
13	C	27	A	41	B	55	A	69	E		
14	B	28	D	42	A	56	E	70	C		

Section V
Practice in answering examination questions

This section of the book is designed to give you practice in answering different types of examination questions. It is therefore arranged in three groupings: multiple choice questions, questions requiring short answers and longer questions. Each grouping has been subdivided according to whether the questions are more suitable for the GCE 'O' level, SCE or CSE examination. GCE candidates, however, may benefit from trying all the questions.

MULTIPLE CHOICE QUESTIONS (answers on page 153)

GCE questions

Type 1

Each question below has five possible answers labelled A, B, C, D, E. For each numbered question select the one most appropriate answer. Within the group of questions each answer may be used once, more than once or not at all.

Questions 1–4

The following are five physical quantities:

A force   D acceleration
B power   E work
C pressure

Which one of the above

1 involves the concept of area?
2 has the same units as weight?
3 has the same units as energy?
4 could be expressed in newton metre per second?

(University of London)

Questions 5–7

The following are associated with atoms or molecules:

A electrons   D neutrons
B ions   E protons
C isotopes

Which of the above

5 form cathode rays?
6 are either positively or negatively charged?
7 are the nuclei of hydrogen atoms?

(University of London)

Type 2

Each of the following questions or incomplete statements is followed by five suggested answers labelled A, B, C, D, E. Select the best answer in each case.

8 Two forces of magnitude 8N and 6N act on the same body. The angle between the directions of the forces is 90°. The magnitude of the resultant of the forces is

A 7N   D 24N
B 10N   E 48N
C 14N

(University of London)

9 The efficiency of a machine is equal to

A $\dfrac{\text{Mechanical Advantage}}{\text{Velocity Ratio}}$

B Mechanical Advantage $\times$ Velocity Ratio

C $\dfrac{\text{Load}}{\text{Effort}}$

D Load $\times$ Effort

E $\dfrac{\text{Distance moved by Effort}}{\text{Distance moved by Load}}$

(University of London)

10 Hot water at 100°C is added to 300 g of water at 0°C until the final temperature is 40°C. The mass of hot water added must be at least

A 60 g   D 180 g

B 75 g   E 200 g

C 120 g   *(University of London)*

11 The fixed point used for calibrating thermometers at 100°C specifies the pressure at which this should be done. This is because changes of pressure alter the

A specific latent heat of vaporisation of water;

B specific heat capacity of water;

C boiling point of water;

D amount that mercury expands;

E density of mercury.   *(University of London)*

12 Which of the following examples of electromagnetic radiation has the shortest wavelength?

A Radio waves   D Visible light

B Infra-red rays   E X-rays

C Ultra-violet rays   *(University of London)*

13 Which of the following does NOT apply to sound waves?

A They transmit energy.

B They result from vibrations.

C They are propagated by a series of compressions and rarefactions.

D They travel fastest in a vacuum.

E They can be diffracted.   *(University of London)*

14 Fuses are available which melt when the current through them exceeds 2, 5, 10, 15, or 30 A. Which one of these would be the most suitable for a circuit in which an electric fire rated at 2.5 kW is to be connected, if the supply voltage is 240 V?

A The 2 A fuse   D The 15 A fuse

B The 5 A fuse   E The 30 A fuse

C The 10 A fuse   *(University of London)*

15 By thermionic emission is meant the emission by an incandescent filament of

A electrons   D ions

B heat   E light

C infra-red radiation   *(University of London)*

Type 3

For each of the questions below, one or more of the responses given are correct. Decide which of the responses is (are) correct. Then choose

A if 1, 2 and 3 are all correct;

B if 1 and 2 only are correct;

C if 2 and 3 only are correct;

D if 1 only is correct;

E if 3 only is correct.

16 A force P is resolved in to two components which are at right angles to each other. The two components have

1 magnitudes which are less than the magnitude of P;

2 directions which are both at an angle of less than 90° to the direction of P;

3 magnitudes which if added together give a value greater than the magnitude of P.

                                                               (University of London)

17 A lever is a device which can enable
   1 a small force to be changed into a large force;
   2 a small movement to be converted into a large movement;
   3 the direction of a force to be changed.   *(University of London)*

18 A hydraulic braking system is commonly used in cars because liquids
   1 find their own level in the tubes;
   2 transmit pressure in all directions;
   3 are almost incompressible.   *(University of London)*

19 Which of the following types of electromagnetic radiation may have a higher frequency than X-rays?
   1 Gamma radiation
   2 Ultra-violet radiation
   3 Infra-red radiation   *(University of London)*

20 A power supply is labelled 12 V d.c., but the positive and negative markings on the terminals have been omitted. You could show which terminal was positive with a circuit using the
   1 heat produced in a coil of resistance wire;
   2 magnetic field in a coil of copper wire;
   3 electrolysis of copper sulphate solution.

   (University of London)

21 Two resistors of 4 Ω each may be connected together in such a way as to produce a combined resistance of
   1 1 Ω
   2 2 Ω
   3 8 Ω   *(University of London)*

22 The figure below shows how a given current varies with time.

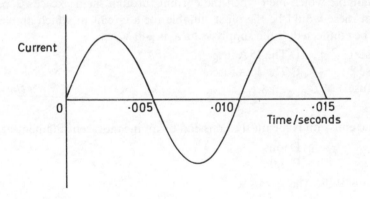

   Which of the following statements about the current is (are) correct?
   1 Three complete cycles of the current are shown;
   2 Such a current cannot be measured with a moving-coil ammeter;
   3 The frequency of the current is 100 hertz.

   (University of London)

SCE questions

Type 2

Each of the following questions or incomplete statements is followed by five suggested answers labelled A, B, C, D, E. Select the best answer in each case.

23 A ball is projected horizontally off a ledge at 10 ms⁻¹. Neglecting air friction, the horizontal component of its velocity two seconds after leaving the ledge will be
   A 20 ms⁻¹   D 1 ms⁻¹
   B 10 ms⁻¹   E 0.5 ms⁻¹
   C 5 ms⁻¹   *(Scottish Certificate of Education Examination Board)*

24 The gravitational field strength on Earth is 10 N kg^{-1}, and on Mars it is is 4 N kg^{-1}. A space probe has a mass of 100 kg on Earth. On Mars

A its mass and weight will have decreased;
B its mass will be smaller, but its weight will be the same;
C its mass will be smaller, but its weight will have increased;
D its mass will be the same, but its weight will have increased;
E its mass will be the same, but its weight will have decreased.

(Scottish Certificate of Education Examination Board)

25 In a supermarket a lady notices that a loaded trolley is difficult to start and difficult to stop. The property of the loaded trolley which accounts for both these observations is its

A friction   D density
B energy   E inertia
C volume   *(Scottish Certificate of Education Examination Board)*

26 During an inelastic collision

A momentum is conserved, but not kinetic energy;
B neither momentum nor kinetic energy is conserved;
C kinetic energy is conserved, but not momentum;
D momentum and kinetic energy are conserved;
E kinetic energy is converted into momentum.

(Scottish Certificate of Education Examination Board)

27 Which physical quantity is a vector?

A mass   D time
B energy   E power
C force   *(Scottish Certificate of Education Examination Board)*

28 Assuming that no heat energy is lost to the surroundings, a temperature of 50°C could be obtained by mixing

A equal masses of ice and boiling water;
B equal masses of ice and steam;
C equal masses of steam and water at 0°C;
D equal masses of water at 25°C and 65°C;
E equal masses of water at 20°C and 80°C.

(Scottish Certificate of Education Examination Board)

29 A burn from steam at 100°C can be worse than a burn from water at 100°C because the steam

A is less dense than water;
B is hotter than the water;
C hits your skin with greater force;
D contains more energy than water;
E has a greater specific heat capacity than water.

(Scottish Certificate of Education Examination Board)

30 A disc which is mounted on the shaft of an electric motor has a single mark on it. The disc is rotated at 50 revolutions per second. When it is illuminated by a stroboscope flashing at 100 flashes per second what will an observer see?

A One stationary mark
B Two stationary marks
C One stationary mark every second flash
D One mark rotating slowly forward
E One mark rotating slowly backward

(Scottish Certificate of Education Examination Board)

31 The bending of waves round an obstacle is known as

A diffraction   D dispersion
B interference   E refraction
C reflection

(Scottish Certificate of Education Examination Board)

32 An electric convection heater contains an element to provide heat and a lamp to provide a glow. One day the lamp stops working but the element continues to work normally. A possible explanation is that the lamp and the element are connected in

A series and the fuse in the plug has 'blown';
B parallel and the fuse in the plug has 'blown';
C series and the lamp filament has broken;
D parallel and the lamp filament has broken;
E series and the lamp has worked loose.

(Scottish Certificate of Education Examination Board)

33 The smallest resistance of a piece of electrical apparatus which may be safely supplied through a 5 A plug on 250 V mains is

A $0.02\,\Omega$ D $50\,\Omega$
B $0.1\,\Omega$ E $1250\,\Omega$
C $10\,\Omega$

(Scottish Certificate of Education Examination Board)

CSE questions

Type 1

Each question below has five possible answers labelled A, B, C, D, E. For each numbered question select the one most appropriate answer. Within the group of questions each answer may be used once, more than once or not at all.

Questions 34–37

The following units are used in Mechanics:

A kilograms per cubic metre (kg/m^3)   D newton (N)
B metre (m)   E newton per square metre (N/m^2)
C joule (J)

Which unit is used to measure the following:

34 energy
35 weight
36 pressure
37 density

(Associated Lancs Schools Examining Board)

Questions 38–41

The following are types of electromagnetic radiation:

A radio waves   D infra-red radiation
B ultra-violet radiation   E X-rays
C visible light

Which type of radiation

38 passes through a thin sheet of lead;
39 causes suntan;
40 is given out from an electric fire;
41 has the longest wavelength?

(Associated Lancs Schools Examining Board)

Type 2

Each of the following questions or incomplete statements is followed by five suggested answers labelled A, B, C, D, E. Select the best answer in each case.

42 Velocity is different from speed because it

A is measured in metric units;
B means distance divided by time;

C is measured in a specified direction;
D is measured in metres per second;
E is measured horizontally. *(Associated Lancs Schools Examining Board)*

43 A body accelerates from rest at 4 m/s² for 5 seconds.
Its average speed in metres per second is
A 0.8 D 10
B 1.2 E 11
C 9 *(East Anglian Examinations Board)*

44 A 10 000 kilogram lunar probe rocket is accelerated by the Moon's gravitational field at 5 metre/second² as it nears the Moon's surface. What force does the Moon exert on the rocket?
A zero D 50 000 newton
B 2000 newton E 100 000 newton
C 10 000 newton *(Met Regional Examinations Board)*

45 The acceleration would be 4 metre/second² if
A a resultant force of 2 newton acts on a mass of 2 kg;
B a resultant force of 4 newton acts on a mass of 16 kg;
C a resultant force of 8 newton acts on a mass of 2 kg;
D a resultant force of 2 newton acts on a mass of 8 kg.

(Met Regional Examinations Board)

46 The mass of a moving body multiplied by its velocity is measuring the body's
A inertia D momentum
B weight E energy
C acceleration *(South Regional Examinations Board)*

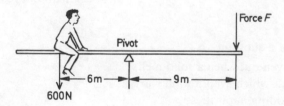

47 A boy weighing 600 newton sits 6 metres away from the pivot of a see-saw, as shown above. What force F, 9 metres from the pivot, is needed to balance the see-saw?
A 300 newton D 600 newton
B 400 newton E 900 newton
C 450 newton *(Met Regional Examinations Board)*

48 A machine is a device which normally
A increases power; D magnifies forces;
B saves time; E creates forces.
C saves energy; *(South Regional Examinations Board)*

49 A stone has a mass of 480 g and a volume of 160 cm³. Its density is
A 76 800 g/cm³ D 3 g/cm³
B 540 g/cm³ E 0.3 g/cm³
C 320 g/cm³ *(Associated Lancs Schools Examining Board)*

50 A car's tyres are inflated so that an area of 50 cm² of each is in contact with the ground. If the mass of the car is 1600 kg and is evenly distributed between the four tyres, what is the pressure exerted on the ground by each one? g = N/kg

A 80 N/cm²   D 320 N/cm²
B 160 N/cm²   E 500 N/cm²
C 200 N/cm²

(South Regional Examinations Board)

51 Which of the following describes particles in a solid?

A Close together and stationary
B Close together and vibrating
C Close together and moving around at random
D Far apart and stationary
E Far apart and moving around at random

(Met Regional Examinations Board)

52 If the absolute temperature of a fixed mass of gas is doubled at constant pressure, its volume will

A increase by 100%;   D decrease by 100%;
B decrease by 50%;   E remain the same.
C increase by 50%;

(South Regional Examinations Board)

53 Five blocks of metal each of mass 1 kg have 1000 J of energy supplied to them by an immersion heater. Which of the blocks shows the greatest temperature rise?

A iron (specific heat capacity 460 J/kg°C)
B copper (specific heat capacity 400 J/kg°C)
C brass (specific heat capacity 380 J/kg°C)
D lead (specific heat capacity 140 J/kg°C)
E aluminium (specific heat capacity 900 J/kg°C)

(Associated Lancs Schools Examining Board)

54 The melting point of a substance is

A the latent heat given out when a solid melts;
B the temperature at which evaporation starts;
C not affected by adding impurities;
D affected by changes in pressure;
E not affected by changes in pressure.

(Associated Lancs Schools Examining Board)

55 A hand feels cold when a little ether is put on it. Which *one* of the following reasons is correct?

A the hand is at a greater temperature than the ether;
B ether insulates the hand from air;
C ether evaporates rapidly and absorbs heat from the hand;
D ether evaporates and forms a layer over the hand;
E ether is at a greater temperature than the hand.

(Associated Lancs Schools Examining Board)

56 When more molecules of a liquid return to it than escape from it, an observer would say that the substance was

A condensing   D evaporating
B conducting   E radiating
C diffusing

(Met Regional Examinations Board)

57 Double glazing improves the heat insulation of houses because

A glass of double thickness does not conduct heat;
B radiation will not pass through two sheets of glass;
C the air trapped between the glass is a bad conductor of heat;
D convection currents between the glass sheets are restricted.

(Met Regional Examinations Board)

58 Convection occurs

A only in solids;
B only in liquids;
C only in gases;
D in solids and liquids;
E in liquids and gases.

(Associated Lancs Schools Examining Board)

59 A piece of aluminium leaf is placed near the palm of your hand. Immediately the hand feels warmer because

A the leaf reflects the heat radiated by your hand;
B the leaf absorbs heat radiated from the surroundings;
C the leaf radiates well;
D the leaf keeps the cooling air away;
E the leaf is a good absorber of radiation.

(South Regional Examinations Board)

60 Which one of the following statements about the image produced in a pinhole camera is correct?

A The image is bigger if the object is further away;
B The image is smaller if the screen is nearer the pinhole;
C The image is brighter if the object is further away;
D The image is sharper if the pinhole is made bigger;
E The image is bigger if the object is brighter.

(South Regional Examinations Board)

61 When you look at yourself in a mirror you see an image of yourself. The image is

A on the surface of the mirror;
B a real image behind the mirror;
C an inverted virtual image;
D caused by rays behind the mirror;
E a virtual image behind the mirror.

(South Regional Examinations Board)

62 When a concave mirror is held close to your face you see an image which is

A smaller and inverted;
B larger and inverted;
C smaller and upright;
D larger and upright;
E the same size and upright.

(Associated Lancs Schools Examining Board)

63 The image formed on the retina of the human eye is

A magnified and erect;
B diminished and inverted;
C life-size and virtual;
D diminished and erect;
E magnified and inverted.

(Yorkshire Regional Examinations Board)

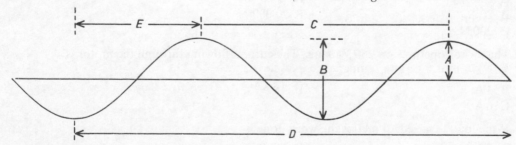

The diagram shows a wave. Which letter shows

64 the wavelength?
65 the amplitude?

(Associated Lancs Schools Examining Board)

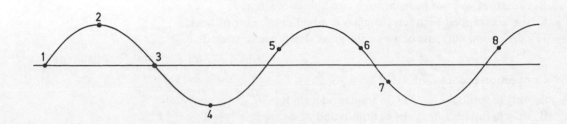

66 The above figure shows a transverse wave. Which two points are in phase?

A 1 and 3   D 5 and 8

B 2 and 4   E 7 and 8

C 5 and 6

(Met Regional Examinations Board)

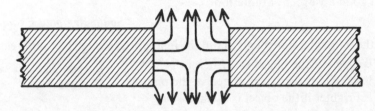

67 The magnetic field pattern shown in the above diagram is produced by bringing together

A N-pole and S-pole   D S-pole and N-pole

B N-pole and N-pole   E S-pole and unmagnetised bar

C S-pole and S-pole

(Associated Lancs Schools Examining Board)

68 In an electric circuit, the number of joules of energy supplied to each coulomb is measured by

A a wattmeter   D an ammeter

B a voltameter   E a voltmeter

C a ratemeter

(South Regional Examinations Board)

69 The current flowing through a 12 V 24 W car headlamp operating on a 6 V supply would most probably be

A 0.5 A   D 2 A

B 1 A   E 4 A

C 1.2 A

(South Regional Examinations Board)

Questions 70–72

A 2 kW electric motor (100% efficient) is used to operate a winding engine.

70 If it raises the load 1 metre in 1 second the load is

A 2 N   D 2000 N

B 20 N   E 20 000 N

C 200 N

71 The motor operates on 250 V mains. The current flowing through the motor is

A 2 A   D 8 A

B 4 A   E 10 A

C 6 A

72 If the load is raised 10 metres the work done is

A 2 J   D 2000 J

B 20 J   E 20 000 J

C 200 J

(Associated Lancs Schools Examining Board)

73 For which one of the following is an alternating current essential in its operation?

A An electromagnet   D An electric lamp
B A transformer   E An electric fire
C A galvanometer

(South Regional Examinations Board)

Questions 74 and 75
A step down transformer has a turns ratio of 6:1. A 240 V alternating current supply at a frequency of 50 Hz is connected to the primary coil.

74 The frequency of the alternating current output in hertz is

A 0.2   C 50
B 10   D 500

75 The voltage of the alternating current output is

A 40   C 240
B 50   D 1440

(East Anglian Examinations Board)

76 A particle with a mass of 1 atomic unit and a charge of $+1$ is called

A an electron   D a proton
B a gamma particle   E an alpha particle
C a positron

(Yorkshire Regional Examinations Board)

77 Which one of the following quantities in an atom represents the mass number of the atom?

A The number of protons plus the number of neutrons.
B The number of protons plus the number of electrons.
C The number of electrons.
D The number of protons.
E The number of neutrons.

(South Regional Examinations Board)

78 When an atom loses an electron it becomes

A a negative ion
B a neutron
C a positive ion
D an alpha particle
E a proton

(South Regional Examinations Board)

ANSWERS TO MULTIPLE CHOICE QUESTIONS

GCE answers

1	C	5	A	9	A	13	D	17	A	21	C
2	A	6	B	10	E	14	D	18	C	22	C
3	E	7	E	11	C	15	A	19	D		
4	B	8	B	12	E	16	A	20	C		

SCE answers

23	B	25	E	27	C	29	D	31	A	33	D
24	E	26	A	28	E	30	B	32	D		

CSE answers

34	C	42	C	50	A	58	E	66	D	74	C
35	D	43	D	51	B	59	A	67	B	75	A
36	E	44	D	52	A	60	B	68	E	76	D
37	A	45	C	53	D	61	E	69	B	77	A
38	E	46	D	54	D	62	D	70	D	78	C
39	B	47	B	55	C	63	B	71	D		
40	D	48	D	56	A	64	C	72	E		
41	A	49	D	57	C	65	A	73	B		

QUESTIONS REQUIRING SHORT ANSWERS (answers on page 159)

Questions of the short-answer type are used by most of the Examination Boards. The answer can be a single word, a phrase, or a few sentences. Some Boards leave a space for the student's answer. This is often an indication of the length of answer required. Other Boards allocate marks, which can also indicate length of answer and the amount of time you should spend on each part of the question.

GCE questions

1 The diagram below is a graph of the journey made by a train. Its speed is measured in metres per second (m/s) and time is measured in seconds (s).

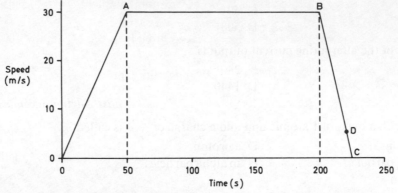

(a) What is the maximum speed of the train?

(b) What is the train doing when its motion is represented by the line *AB*?

(c) What is the train doing when its motion is represented by the line *BC*?

(d) Which of the points *O, A, B* or *C* represents the stage at which the brakes are applied?

(e) The line *BC* is steeper than the line *OA*. What does this tell you about the rates at which the train speeds up and slows down?

(f) Calculate how far the train travelled between the stages in its journey represented by the points *O* and *A*.

2 An object is placed 15 cm in front of (i) a plane mirror, (ii) a concave mirror of radius 60 cm, and (iii) a convex mirror of radius 60 cm. Draw scale diagrams to show where the images are produced in each case. State the nature and position of each image.

3 What type of lens should be used in a compound microscope

(i) as the eyepiece?

(ii) as the objective?

In each case state if the lens should have a long or short focal length.

4 A detector shows that the activity of a radioactive sample falls from 160 units to 20 units in 15 minutes. Determine the half-life of the sample, explaining your calculation.

SCE questions

5 A vehicle is fired along a frictionless air track by a stretched elastic band. A card, of length 14 cm, is attached to the vehicle. A clock measures the time taken for this card to pass through a light beam at point A.

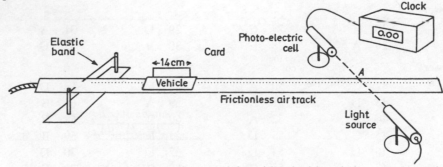

The experiment is repeated using two, three and four identical bands in turn. In each case the bands are stretched back the same distance before the vehicle is released. The results are given in the table.

Number of elastic bands	Time recorded on clock (s)	Speed of vehicle at A (m s⁻¹)	Kinetic Energy of vehicle at A (J)
1	0.14		
2	0.10		
3	0.08		
4	0.07		

(a) The total mass of vehicle and card is 1.0 kg. Copy and complete the above table.

(**5 marks**)

(b) Use the results to show that kinetic energy varies directly as the number of elastic bands.

(**3 marks**)

(c) Explain why the elastic bands were stretched back the same distance in each case.

(**2 marks**)

(**10 marks**)

(*Scottish Examination Board*)

6 In an experiment to find how the resistance of a coil of wire varies with temperature, the following results were obtained.

Temperature (°C)	Resistance (Ω)
31.0	53.5
41.5	55.5
51.0	57.0
61.5	59.0
71.0	60.5
81.0	62.0

(a) (i) Draw a voltmeter-ammeter circuit which could be used to measure the resistance of the coil.
(ii) Describe how this circuit can be used to obtain the above results. (**5 marks**)

(b) Using square-ruled paper plot a graph of the results and from it find the resistance of the coil at 0°C.

(**5 marks**)

(**10 marks**)

(*Scottish Examination Board*)

7 A loaded mine car of total mass 5 tonnes (5×10^3 kg) is being hauled up an incline to the surface when the cable breaks and the car runs back down the incline.

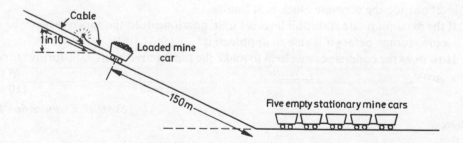

At the bottom of the incline the car collides with a row of five stationary empty cars, each of mass 1 tonne. The cars automatically couple together and move off with a speed of 4 m s⁻¹.

(a) Use the law of conservation of momentum to calculate the speed of the loaded car just before impact. **(3 marks)**

(b) Calculate the kinetic energy of the loaded car just before impact. **(2 marks)**

(c) For every 10 m it is hauled up the incline, the car rises a vertical height of 1 m. If the cable broke when the car was 150 m up the incline, what was its potential energy at that point? **(3 marks)**

(d) Account for any difference between the answers to (b) and (c). **(2 marks)**

(10 marks)

(Scottish Examination Board)

8 An immersion heater is placed in a vacuum flask containing a quantity of liquid and switched on.

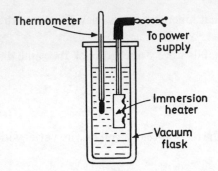

The temperature of the liquid is recorded every 50 seconds and the following table of results is obtained.

Time (s)	0	50	100	150	200	250
Temperature (°C)	20	51	82	114	130	130

(a) Plot a graph of these results. **(3 marks)**

(b) Estimate from the graph as accurately as you can
  (i) the boiling point of the liquid;
  (ii) the time for the liquid to reach its boiling point. **(3 marks)**

(c) Explain how the information on the graph may be used to calculate the specific heat capacity of the liquid, listing any extra data needed to do this. **(4 marks)**

(10 marks)

(Scottish Examination Board)

9 A nuclear power station supplies an average power of 150 MW to the National Grid. The power station's reactor consumes 0.3 kg of uranium fuel per hour.

(a) (i) Calculate the energy supplied to the National Grid in 1 hour.
  (ii) If the energy extracted from 1 kg of uranium fuel in the nuclear reactor is 5.4×10^{12} J, calculate the efficiency of the power station. **(5 marks)**

(b) (i) Part of the radioactive waste from the reactor has a half-life of 30 years. A drum of this waste material is encased in a concrete block before storage. The measured radiation level outside the concrete block is 512 units.
    If the maximum safe allowable level is 1 unit, how long must the concrete block be kept in secure storage before it is safe to approach it?
  (ii) How does the concrete casing help to make the radioactive waste safe during storage?

(5 marks)

(10 marks)

(Scottish Examination Board)

CSE questions

10 A page of newspaper is crumpled and rolled into a ball. Explain why this page would reach the ground before an exactly similar, but uncrumpled, page if they were released together from the same height.

(East Anglian Examinations Board (N))

11 The figure is a graph of height against time when a ball is dropped from position *A* on to the floor at *B*, 1 m below it. The ball bounces and reaches a height *C*, which is 0.5 m above the floor, and falls to the floor again.

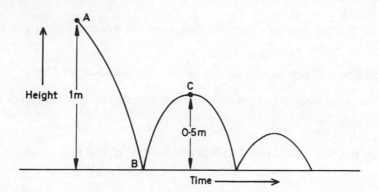

(*a*) Supposing that the ball has 1 J of potential energy at *A*, say what its kinetic energy will be
   (i) just before it hits *B*
   (ii) just after it has moved away from *B*.
(*b*) What will be its potential energy at *C*?
(*c*) Why does the ball only reach *C* and not 1 m above the floor?

(Middlesex Regional Examining Board)

12 A boy weighing 400 N runs up a flight of steps 10 m high in 5 seconds.
(*a*) How much work does he do?
(*b*) What power is generated?

(Associated Lancs Schools Examining Board)

13 A windpump raises 2500 N of water to the surface from 6 m below ground level in 300 seconds.
(*a*) Find the work done by the pump in that time.
(*b*) Calculate the power output of the pump.

(East Anglian Examinations Board(N))

14 A 50 kg load is pulled up the slope from *A* to *B* through a distance of 8 m as indicated in the diagram below, by a force of 500 N.
(i) What is the velocity ratio of the system?
(ii) What is the efficiency of the system?

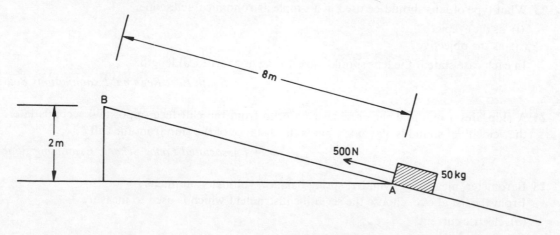

(South East Regional Examinations Board)

15 (*a*) Calculate the surface area of a cube of side 2 cm.
(*b*) Calculate the volume of the same cube.
(*c*) Find its mass if it has a density of 7 g/cm³.

(East Anglian Examinations Board (N))

16 A concrete paving slab measures 0.6 m × 0.5 m × 0.05 m.
  (*a*) What is its volume?
  (*b*) Taking the density of concrete to be 2400 kg/m³, what is
    (i) the mass of the slab;
    (ii) its weight in newtons?
  (*c*) What is the average pressure the slab exerts on the ground beneath it when it is laid horizontally?

17 On what factors does the pressure under the surface of a liquid depend?

18 A rectangular block of wood has a mass of 32 g and measures 5 cm long, 4 cm wide and 2 cm thick.
  (i) What is its density?
  (ii) Explain why the block would float at different levels if placed in sea water and fresh water.
  (South East Regional Examinations Board)

19 Complete the following sentences:
  (*a*) Heat travels by _____ in solids.
  (*b*) Heat travels mainly by _____ in liquids and gases.
  (*c*) Heat travels mainly by _____ in space.
  (Associated Lancs Schools Examining Board)

20 A ship's echo sounder receives a pulse of sound 3 seconds after it was emitted. Find the depth of the sea under the ship given that the velocity of sound in water is 1500 m/s.
  (East Anglian Examinations Board(N))

21 Fill in the blank spaces with the correct word(s) in the following passage.
  A person suffers from long sight because his eye lens is too _____. Because of this the image is brought to focus _____ the retina. This fault may be remedied by using spectacles with a _____ lens so that the image is now formed _____ the retina.
  (Middlesex Regional Examining Board)

22 Fill in the blank spaces with the correct word(s) in the following passage.
  A person suffers from short sight because his eyeball is too _____. Because of this the image is brought to focus _____. the retina. This fault may be remedied by using spectacles with a _____ lens so that the image is now formed _____ the retina.

23 What type of lens should be used in a simple astronomical telescope
  (i) as the eyepiece?
  (ii) as the objective?
  In each case state if the lens should have a long or short focal length.
  (South East Regional Examinations Board)

24 A ship fires a gun and the crew hear an echo from the cliff face opposite 4 seconds later. If the velocity of sound is 300 m/s, what is the distance of the ship from the cliff?
  (Associated Lancs Schools Examining Board)

25 Barometer, measuring cylinder, spring balance, voltmeter, ammeter.
  From the list above, choose the scientific instrument which is used to measure
  (*a*) electric current;
  (*b*) atmospheric pressure;
  (*c*) the electromotive force of a battery.

26 Calculate the total value of a 4 ohm and 8 ohm resistor
  (*a*) when joined in series,
  (*b*) when joined in parallel.

Answers to short answer questions
GCE answers

1 (a) 30 m/s
   (b) travelling at a constant speed
   (c) decelerating uniformly

(d) B
(e) The train slows down at a greater rate than it speeds up
(f) 750 m.

2 (i) 15 cm behind the mirror surface. Erect, virtual and the same size as the object.
  (ii) 30 cm behind the mirror surface. Erect, virtual and magnified.
  (iii) 10 cm behind the mirror surface. Erect, virtual and diminished.

3 (i) A converging lens of short focal length.
  (ii) A converging lens of short focal length.

4 5 minutes. In 15 minutes the activity falls to 1/8th of its original value; that is it halves itself three times. It therefore halves itself once in 5 minutes.

SCE answers

5 Speeds of the vehicle are 1 m/s, 1.4 m/s, 1.75 m/s and 2 m/s. Kinetic energies of the vehicle are 0.5 J, 0.98 J, 1.5 J and 2 J. The elastic bands are stretched back the same distance in each case so that each time each elastic band stores the same amount of potential energy. When this potential energy is transformed into kinetic energy, the quantity transformed is thus proportional to the number of bands.

6 (a) (i) The ammeter, the coil and some cells are connected in series. The voltmeter is connected in parallel with the coil.
    (ii) The coil should be connected into the circuit in such a way that it can be immersed in a beaker of water. The water is heated to the required temperatures and, at each temperature, readings of the thermometer (in the water), ammeter and voltmeter are taken. The resistance is calculated from the ammeter and voltmeter readings.

7 (a) 8 m/s (b) 1.6×10^5 J (c) 7.5×10^5 J (d) Energy transformed to heat in overcoming frictional forces.

8 (b) (i) 130° (ii) 200 s
  (c) Each second the energy supplied by the heater equals the heat gained by the liquid provided none is lost. Up to 100°C the graph is a straight line indicating that no energy is lost.
    Power of the heater = mass of liquid × specific heat capacity × temperature rise/time
    Temperature rise/time is the gradient of the straight line section of the graph. The power of the heater and the mass of the liquid need to be known.

9 (a) (i) 5.4×10^{11} J   (ii) 33.3%
  (b) (i) 270 years   (ii) It absorbs much of the radiation given out by the waste.

CSE answers

10 The crumpled page experiences a smaller frictional force due to air resistance than the uncrumpled page.

11 (a) (i) 1 J   (ii) 0.5 J
   (b) 0.5 J
   (c) The collision the ball makes with the floor at B is inelastic; that is, some kinetic energy is transformed into heat and sound.

12 (a) 4000 J   (b) 800 W
13 (a) 15 000 J   (b) 50 W
14 (i) 4   (ii) 25%
15 (a) 24 cm²   (b) 8 cm³   (c) 56 g
16 (a) 0.015 m³   (b) (i) 36 kg   (ii) 360 N   (c) 1200 N/m²

160

17 The depth and density of the liquid and the value of the acceleration due to gravity.
18 (i) 0.8 g/cm³
    (ii) The block floats at a level such that it displaces its own weight of liquid. Sea water has a greater density than fresh water and so less volume has to be displaced for the same weight.
19 (a) Conduction   (b) Convection   (c) Radiation
20 2250 m
21 weak, behind, converging (convex), on.
22 long, in front of, diverging (concave), on.
23 (i) A converging lens of short focal length.
    (ii) A converging lens of long focal length.
24 600 m
25 (a) ammeter   (b) barometer   (c) voltmeter
26 (a) 12 ohms   (b) $\frac{32}{12}$ ohms

LONGER QUESTIONS

There follows a selection of questions, mainly reprinted from past examination papers of the GCE, SCE and CSE Boards. The Examination Board from which each question comes is indicated at the end of the question.

The questions are numbered according to the relevant unit in the text. However, some questions involve work covered in more than one unit, and in these cases the questions occur in the last relevant unit. Such questions thus provide an opportunity for revision of work in an earlier unit.

Each question is followed by a few hints about how to answer it (printed in red type). These hints are NOT intended as 'model answers'; they are merely meant to indicate some of the important physical principles involved in the question. It is hoped that the hints will aid your understanding of the topic.

Answers to the numerical parts of questions are given after the last question (see pages 216-218).

GCE questions
Unit 2

1 The following readings were taken in an experiment to determine the acceleration of free fall g.

Height of fall from rest h(cm)	200	180	160	140	120	100
Time taken t(s)	0.64	0.61	0.54	0.53	0.50	0.45
t^2(s²)		0.37	0.29	0.28	0.25	

Complete the table by inserting the two missing values of t^2 corrected to two significant figures. Plot a graph of h against t^2.

Which pair of readings appears to have been measured or recorded incorrectly?

h = ... t = ...

Ignoring that pair of readings, draw the best straight line through the other points on your graph.

What is the gradient of the straight line obtained?............

Hence determine an approximate value for g stating clearly the units of your answer.

g = ...   units............   *(Associated Examining Board)*

The points give a good straight line through the origin if the point for $h = 160$ cm is ignored as being too high.

The gradient is found by choosing a point towards the top of the straight line and dividing the value of h by the value of t^2. Choose a point with a convenient value of t^2, *e.g.* 0.4.

The value of the gradient is 490 cm/s² approx.

For free fall, $g = \dfrac{2h}{t^2}$ (unit 2.8). Thus $g = 980$ cm/s² or 9.80 m/s².

Unit 3

2 (*a*) Define the *newton*.

Describe an experiment to investigate the relationship between the force applied to a trolley and the acceleration produced in the trolley.

(*b*) The speed of a car of mass 500 kg is 5 metres/second when the brakes are applied. The car comes to rest in 10 metres. Find (i) the average retardation; (ii) the average braking force.

(*a*) See unit 3.2.

(*b*) Use $v^2 = u^2 + 2as$ to find a, the average retardation

$$\left(\text{or find } t \text{ from } s = \frac{1}{2}t(u + v) \text{ and hence } a = \frac{v - u}{t} \right).$$

$F = ma$ then gives the average braking force.

3 What is meant by *acceleration*? What evidence have you for believing that a moving body will proceed at a constant speed in a straight line indefinitely unless it is acted on by a resultant force?

A cyclist of mass 50 kg riding a cycle of mass 10 kg raises his speed from 2 m/s to 5 m/s by accelerating uniformly for 6 s while travelling due north along a horizontal road. What horizontal force is exerted on the machine by the road, and in what direction? Does the cyclist exert any force on the machine, and, if so, in what direction?

How far does he travel in 6 seconds? (*Oxford and Cambridge Schools Examination Board*)

Acceleration is the rate of change of velocity, that is it is calculated by dividing the change by the time taken. If the acceleration is not uniform the change considered must be over a short time period. Evidence for straight line motion comes from CO_2 pucks moving on a clean smooth horizontal surface or from an air table. If gravity is ignored evidence is obtained when the centripetal force is removed from a body undergoing circular motion (*e.g.* string breaks when a stone is being whirled horizontally).

The acceleration ($0.5\,\text{m/s}^2$) of the cyclist and machine are calculated as indicated above. The road exerts a force on the machine, in a northerly direction, to accelerate both the machine and the cyclist. This force is calculated using $F = ma$ where m is the combined mass. The cyclist exerts a force perpendicular to the pedals.

The distance travelled is the average speed (3.5 m/s) multiplied by the time (6 s).

4 A boy holds the end, *O*, of a rubber cord from which hangs a solid metal ball.

(*a*) When the ball is still, as in Fig. (i), the rubber stretches. Draw a diagram showing the forces acting on the ball. What do you know about these forces?

Fig. (i)

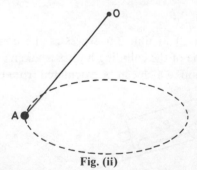

Fig. (ii)

(*b*) In Fig. (ii) the boy whirls the ball in a horizontal circle, keeping his hand still. Draw a diagram showing the forces acting on the ball in position *A*. Explain carefully why the rubber cord stretches more in the situation of Fig. (ii) than that of Fig. (i).

(*c*) Explain why each of the forces in the situation of Fig. (ii) does no work as the ball rotates.

(*d*) If the rubber cord breaks when the ball is at *A*, state clearly the direction in which the ball will move, giving a reason for your answer.

(*e*) At *A* the ball is 1.25 m above the ground. Calculate the time it will take to reach the ground after the cord breaks. (The time is the same as if the ball were to fall vertically from rest.)

(*f*) If at the moment the cord breaks the ball is moving at 1.5 m/s, what will be the horizontal distance from *A* when it reaches the ground?

(University of London)

(*a*) The ball is stationary, thus the forces on it are in equilibrium. Its weight is equal in size but opposite in direction to the upward force exerted by the rubber.

(*b*) The only forces acting on *A* are (i) its weight vertically downwards and (ii) the tension in the cord along *AO*. The upward force, equal to the ball's weight, and the centripetal force acting on the ball, which keeps it moving in a circle, are both provided by the rubber, and together are greater than it provides in (*a*). To provide this greater force the rubber stretches more.

(*c*) In Fig. (ii) because work = force × distance and the ball does not move in the direction of any of the forces, therefore work is not done (see unit 3.7).

(*d*) At any moment the ball is moving horizontally along a tangent to the circle. If the rubber cord breaks the ball will initially travel along that tangent at a constant speed.

(*e*) See units 2.7 and 2.8.

(*f*) During the time it takes to reach the ground the ball travels horizontally at a constant speed of 1.5 m/s. The distance is found by multiplying these two together (time × velocity).

5 What is meant by the principle of conservation of linear momentum?

How would you show that the principle applies to a collision involving one moving trolley and one stationary trolley when the two trolleys join together after the collision?

(Joint Matriculation Board)

In a collision the sum of the separate momenta of all the bodies, in a given direction, is the same afterwards as it is before.

See unit 3.6.

6 What is meant by the *momentum* of a body? State the law of conservation of linear momentum, including the conditions under which it applies.

Describe an experiment which illustrates the validity of the law for simple collisions.

Two bodies of masses 200 kg and 100 kg travel towards each other with velocities of 20 m/s and 25 m/s respectively, and join to form one body on collision.
  (i) What is their common velocity after the collision?
  (ii) In what direction do they move after the collision?

The momentum of a body is the product of its mass and its velocity. It is a measure of how difficult it is to alter the motion of the body. Conservation of linear momentum was discussed in connection with the previous question. For it to apply there must be no other forces acting on the bodies concerned.

See unit 3.6.

Equation (12) in unit 3.6 expresses the conservation law mathematically. The direction of travel of one of the colliding bodies is taken as positive and both bodies have the same velocity after collision, which can be calculated from the equation.

7

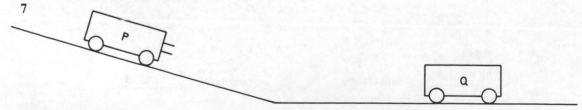

P and *Q* are two freely running trolleys, the mass of *Q* being twice that of *P*. *P* is released from the top of the incline, and *Q* is stationary on the horizontal plane. (All surfaces are frictionless.)
(*a*) Explain how you would find experimentally
  (i) the acceleration of *P* down the incline, and
  (ii) the speed of *P* when it reaches the bottom of the incline.
State clearly the measurements you would take and show how you would use them.

(b) When they collide P and Q stick together. What fraction of P's speed on impact will be the speed of the combined trolleys?

(c) The trolleys are now interchanged and the experiment is repeated, using the same starting positions. What differences, if any, between the two experiments, would you expect in, (iii) the speeds of P and Q as each reaches the horizontal, (iv) the kinetic energies of P and Q as each reaches the horizontal, and (v) the speeds of the combined trolleys?
Give your reasons each time.   *(University of London)*

A length of tickertape is attached to the rear of trolley P. As P accelerates down the slope it tows the tape through a ticker-timer. A histogram is formed from the resulting tape as described in unit 2.8.

(a) (i) Each length of tape equals the distance the trolley travels in 0.1 s. The average velocity during each length of tape (5 dots) is thus found by measuring its length and dividing by 0.1 s.

$$\text{Acceleration } a = \frac{\text{change in velocity}}{\text{time taken}} = \frac{v - u}{t}$$

The change in velocity $(v - u)$ between tape lengths 1 and 6, for example, may be found by subtracting the two velocities concerned. This change in velocity takes place over 25 dots. So $t = 0.5$ s.

(ii) The tape is inspected to find the position of maximum separation of dots just before the collision. A 5-dot length in this region is used to calculate the maximum speed (velocity) of P as described above.

(b) The principle of momentum conservation (unit 3.6) is used to calculate the ratio of the combined speed to P's speed on impact.

(c) (iii) In each case all the trolley's potential energy is converted into kinetic energy. Thus $mgh = \frac{1}{2}mv^2$, where h is the vertical height. The speeds depend only on the values of h and g, which are the same in each case.

(iv) Speeds are the same and KE $= \frac{1}{2}mv^2$. Thus $\text{KE}_1/\text{KE}_2 = m_1/m_2$.

(v) As Q has more mass it has more momentum on impact. Use of the law of momentum conservation gives the speed of the combined trolleys.

8 (a) For a car of mass 2000 kg travelling at 15 m/s, calculate
   (i) Its kinetic energy;
   (ii) Its momentum.

(b) If the car is brought to rest by the brakes with uniform retardation in 12 s, find (i) the acceleration and (ii) the effective retarding force.

(c) Suppose that, instead of applying the brakes, the driver had switched off the engine and allowed the car to run freely up a hill which rises vertically by 1 m for every 10 m travelled along the slope, how far would it have travelled before coming to rest?
                   (Oxford Local Examinations)

(a) (i) KE $= \frac{1}{2}mv^2$
   (ii) Momentum $= mv$

(b) (i) Acceleration $= \dfrac{\text{change in velocity}}{\text{time taken}}$ (-1.25 m/s^2). The negative sign indicates slowing down.

(ii) Force $=$ mass $\times$ acceleration.

(c) The kinetic energy acquired by the car is slowly converted to potential energy as it climbs the slope. When it stops $\frac{1}{2}mv^2 = mgh$ where h is the vertical height reached. The distance travelled along the slope is $10h$. This answer neglects air resistance and friction in the car.

9

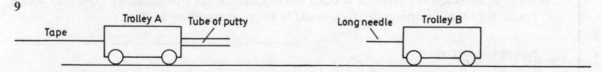

The diagram shows two trolleys, A and B, at rest on a friction-compensated slope. The trolley A, connected to a tape going through a ticker-tape timer, has a tube packed with putty attached to it, and the total mass is 2.0 kg. The trolley B of total mass M has a long needle attached to it, so that the two trolleys stick together when they collide.

Trolley A is given a velocity by means of a sudden push when the ticker-tape timer is working, and the trolleys collide and move along the slope together. The recording on the tape, where the dots occur every 1/10 second, is shown (actual size) at the side of this page. The mark P shows the instant at which the needle hit the putty, and PQ is the length of needle stuck in the putty after the collision. There were two dots between P and Q.

(a) What was the speed of A before the collision?

(b) What was the speed of A after the collision?

(c) The two dots in the section PQ are not shown on the diagram. Copy the section PQ and show on it the likely places for these two dots, assuming that the putty causes a uniform retardation.

(d) How long, in time, did it take for the needle to become stuck in the putty?

(e) What was the average retardation (deceleration) of A?

(f) What was the average force exerted on A during the collision?

(g) What was the average force exerted on B during the collision? Give your reasoning for this answer.

(h) Calculate the mass, M, of B.

(i) Calculate the loss in kinetic energy of A due to the collision and state what becomes of this energy, assuming that the collision makes no sound.

(Oxford and Cambridge Nuffield)

(a) The average spacing between the dots before P is measured. When divided by 0.1 s, this gives the velocity of A before the collision.

(b) As for (a) but using the dots after Q.

(c) The spacing of the dots between P and Q should decrease steadily.

(d) The distance PQ contains three gaps between dots and thus takes 0.3 s.

(e) Retardation $= \dfrac{\text{decrease in speed}}{\text{time taken}}$ and is calculated using the answers to (a), (b) and (c).

(f) Use, Force = mass × acceleration.

(g) The force exerted on B is equal to that exerted on A, but opposite in direction.

(h) Use the answers to (b) and (c) to calculate the acceleration of B. Use $F = ma$ to calculate the mass from this value, and the answer to (g).

(i) Substitute the answers to (a) and (b) in KE $= \frac{1}{2}mv^2$. The loss in kinetic energy becomes heat in the putty. Some kinetic energy is transferred to B.

10

Speed of car	10 m/s	15 m/s	20 m/s	25 m/s	30 m/s
Thinking distance	6 m	9 m		15 m	18 m
Braking distance	8 m	18 m		50 m	72 m
Overall stopping distance	14 m	27 m		65 m	90 m

The table above is similar to that in the Highway Code and shows how the overall stopping distance for a car varies with the speed of the car. The overall stopping distance is the sum of the thinking distance and braking distance.

(a) What do you think is meant by 'thinking distance'?

(b) How does thinking distance depend on the speed of the car?

(c) What is the thinking distance which corresponds to a speed of 20 m/s?

(d) The figure below is a graph of braking distance against the (speed of car)^2. Use it to decide what the overall stopping distance would be for a car travelling at 20 m/s.

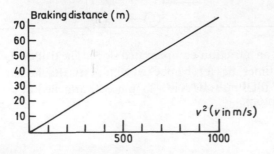

(*e*) Explain why the graph is a straight line through the origin.

(*f*) Outline the energy changes taking place when a car is brought to rest by the application of brakes on a level road.

(*g*) A car could be stopped by running it up a slope.

     If the car were travelling at 20 m/s, to what maximum vertical height, *h*, could it go before coming to rest? Assume that friction can be ignored.

(*h*) Suggest, with reasons for your answer, the minimum safe distance that a car should be driven behind another if both are travelling at 15 m/s.

(Oxford and Cambridge Nuffield)

(*a*) The distance the car travels in the time which elapses between the danger arising and the driver commencing to apply the brakes.

(*b*) The thinking time for the driver is constant. The thinking distance is thus proportional to speed of the car.

(*c*) Use the answer to (*b*) to calculate this.

(*d*) The (speed of car)^2 is 400. Use the graph to find the corresponding braking distance.

(*e*) The (speed of car)^2 is proportional to its kinetic energy. The braking distance of the car will also be proportional to its kinetic energy. It must go through the origin as a car with no speed requires no braking distance.

(*f*) Kinetic energy → heat in the brakes. Some dissipated by air resistance and friction.

(*g*) Ignoring friction, the kinetic energy at the foot of the slope is all converted to potential energy when the car stops. Use $\frac{1}{2}mv^2 = mgh$ to calculate *h*.

(*h*) Strictly speaking only the thinking distance need separate the cars, as they both require the same braking distance. This is rather dangerous as it assumes the car in front will not stop dead in a collision. It also assumes both drivers have the same reaction time. It is safer to leave the overall stopping distance between the cars.

11 Give the meaning of the terms *force*, *work*, *power* and state, in each case, a unit of measurement.

     An electric train, weight 7.2×10^6 N, runs, with its brakes and motors off, down a slope of 1 in 200 (*i.e.* 1 m vertical fall for every 200 m down the track) at a uniform speed of 40 km/hour.

(*a*) What is the component of the weight acting along the track?

(*b*) What can be deduced about the force opposing the motion of the train from the fact that the speed of the train is uniform?

(*c*) At what rate is work being done against the force opposing motion?

     When the train is being driven up the incline at the same speed of 40 km/hour, find

(*d*) the force opposing motion, and

(*e*) the power dissipated in overcoming the force.

(Oxford and Cambridge)

See units 3.1, 3.7 and 3.11.

(*a*) 1/200th.

(*b*) If the speed of the train is uniform there is no resultant force on it. The frictional force opposing motion is thus equal and opposite to the component of weight along the track.

(*c*) Rate of doing work $= \dfrac{\text{force} \times \text{distance moved}}{\text{time taken}} = \text{force} \times \text{velocity}.$

     The rate of doing work is thus found from the answer to (*a*) and the velocity of 40 km/hour (converted to m/s).

(*d*) This force is the sum of the component of weight along the track which the train is overcoming, and the frictional force (these are equal – see (*b*)). This force is thus twice that in (*a*).

(*e*) Twice the answer to (*c*) as the speed is the same.

Unit 4

12 (*a*) What is the difference between a vector quantity and a scalar quantity? Give an example of each.

(*b*) A man on the large flat deck of a ship walks at 5 km/h in a direction perpendicular to the course of the ship. If the ship is travelling due North at 20 km/h, determine, either graphically or by calculation, the true velocity of the man relative to the Earth.

(*c*) State the principle of conservation of momentum.

(d) A trolley of mass 0.50 kg travelling in a straight line at 3.0 m/s collides with and becomes attached to a trolley of mass 1.0 kg travelling in the same straight line at 2.0 m/s. Calculate their common speed after the collision,
   (i) if they were travelling in the same direction;
   (ii) if they were travelling in opposite directions.

(Associated Examining Board)

(a) See unit 4.1.
(b) Resultant of two velocities at right angles = $\sqrt{20^2 + 5^2}$ (unit 4.1)
(c) See unit 3.6. The sum of the momenta of bodies in a given direction before collision is equal to the sum of their momenta in the same direction after collision.
(d) See unit 3.6. Equation (12).

13 (a) A girl can row a boat through still water at 4 km/h. She wishes to row due East across a river 2 km wide and which is flowing due South at 3 km/h.
Either by scale drawing or by calculation, find
  (i) the direction in which she must head the boat;
  (ii) the time she would take to reach the other bank;
  (iii) how far from her destination she would be when she reached the other bank if she mistakenly steered due East.
(b) A bullet of mass 10 g is shot into a block of wood of mass 990 g resting on a smooth horizontal table. If, at the moment of impact, the bullet is travelling horizontally at 300 m/s, calculate the velocity after the impact of the block with the bullet embedded in it. If the two were then struck by a second bullet of mass 20 g travelling horizontally in the same direction at 156 m/s, calculate the common velocity just after the second impact.

(Associated Examining Board)

(a) The girl must row across the river heading slightly up stream, so that the combination of her speed of 4 km/h in this direction and the river's speed of 3 km/h gives a resultant directly across the river.

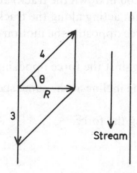

  (i) obtained from scale drawing or sin θ = 3/4;
  (ii) obtain the resultant velocity from drawing, or resultant = $\sqrt{4^2 - 3^2}$ by calculation. Divide the distance across by the resultant velocity.
  (iii) If she heads due East she will cross the river at 4 km/h, that is in half an hour. During this time she will be carried $1\frac{1}{2}$ km down stream.
(b) Use equation (12) in unit 3.6 to apply the conservation of momentum, remembering that the combined mass after collision is 1000 g. Use the same equation again using your answer to the first part, and masses of 1000 g, 20 g, and 1020 g.

14 Explain with the help of a diagram how two forces not in the same straight line may be added.

Show how to resolve a force into two components at right angles.

A uniform rigid rod *AB* of weight 45 N is suspended from a fixed point *C* by strings *AC*, *BC* of equal length. The angles *CAB*, *CBA* are each 13°. When the system is in equilibrium, find, by drawing or calculation,
(a) the tension in each string;
(b) the forces tending to compress the rod due to the tension in the strings.

(Oxford and Cambridge)

See units 4.2 and 4.3.
See unit 4.4.

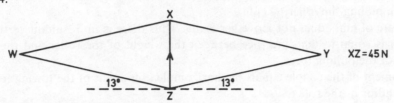

(a) Force represented by length WX or XY.
(b) Force represented by length WY.

Unit 5

15 Explain the expressions *moment of a force about an axis, centre of mass.*

Describe how you would find experimentally, the position of the centre of mass of a flat sheet of wood of irregular outline.

A uniform plank AB of wood 3 m long, weighing 54 N, rests on a knife-edge 0.75 m from B. The end A is supported by a vertical string so that AB is horizontal.

Find (a) the tension T in the string, (b) the force F acting on the knife-edge.

(Oxford and Cambridge)

See the beginning of unit 5 and unit 5.3.
See unit 5.4.

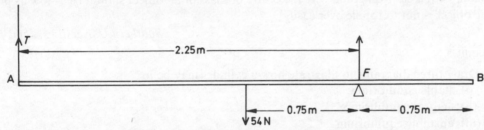

Take moments about the knife-edge, equating the moment due to the weight of the plank acting at a distance of 0.75 m from it to the moment of the string acting at a distance of 2.25 m from it.

The force F acting on the knife-edge is the difference between the weight of the plank and the tension in the string. This is because the total vertical force on the plank must be zero as the plank does not move.

16 The figure represents a uniform roller, of mass 300 kg and radius 0.4 m, which is being pulled along rough horizontal ground at a steady speed by a force of 200 N acting at 30° to the horizontal.

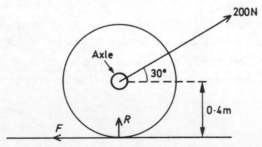

(a) What are the horizontal and vertical components of this force?
(b) What is the frictional force F exerted by the ground?
(c) What is the upward force (reaction) R exerted by the ground?
(d) What is the moment of the couple causing the roller to rotate?
(e) What is the moment of the frictional couple opposing the rotation of the roller? Explain.
(f) Where is the friction in (e) located?
(g) If the roller rolls without slipping explain why no work is done by the applied force against the friction at the ground.

(Oxford Local Examinations)

(*a*) Resolve the force of 200 N into horizontal and vertical components (unit 4.3).

(*b*) As the roller does not slip the frictional force must be equal to the horizontal component of the force pulling the roller.

(*c*) The centre of mass does not move vertically. Thus there is no resultant vertical force. The reaction is equal to the difference between the weight of the roller and the vertical component of the pulling force.

(*d*) The moment of the couple equals the horizontal component of the towing force multiplied by the radius of the roller.

(*e*) As the roller is being pulled at a steady speed the moment of this couple is equal in size and opposite in sense to that in (*d*).

(*f*) Where the roller is in contact with the ground.

(*g*) As the roller does not slip, its surface does not move in the direction of the frictional force.

17 (*a*) Explain what is meant by the *centre of mass* of a body. Describe how you would determine the position of the centre of mass of an irregularly shaped plane sheet of metal.

(*b*) The acceleration due to gravity can be determined by measuring with a stop-watch, the time taken for a body to fall freely from rest through a measured distance. State the probable causes of error in such a determination and suggest ways of reducing the errors.

(Joint Matriculation Board)

(*a*) See units 5.3 and 5.4.

(*b*) Errors arise in starting and stopping the watch, reading it, and measuring the distance fallen. These can be reduced by timing the fall over as great a distance as possible.

18 Explain, with a clear diagram, why the centre of mass of an object should be as low as possible if the object is not to topple over easily.

(Southern Universities' Joint Board)

See unit 5.5.

19 (*a*) Draw three diagrams to illustrate how a cylinder may be in
　　(i) stable equilibrium;
　　(ii) neutral equilibrium;
　　(iii) unstable equilibrium.
State the conditions that produce each type of equilibrium.

(*b*) A body of mass 10 kg, initially travelling with a constant velocity of 7 m/s, is subjected for a period of 6 seconds to a force of 50 N, acting in the direction of motion of the body. Calculate
　　(i) the velocity of the body at the end of the six seconds;
　　(ii) the distance travelled by the body during the six seconds.

(Associated Examining Board)

(*a*)　(i) A small displacement of the centre of mass results in a moment which tends to decrease the displacement. Stable.
　　(ii) A small displacement of the centre of mass results in no moment.
　　(iii) A small displacement of the centre of mass results in a moment which tends to increase the displacement. Unstable.

These positions are achieved respectively by standing the cylinder on one end, on its curved side, and on an edge.

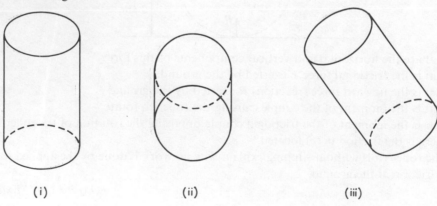

(i)　　　　　　　　(ii)　　　　　　　　(iii)

(b) Use equation (10) in unit 3.5 or use $F = ma$ to calculate the acceleration produced, and then work from the definition of acceleration given in unit 2.5. The distance travelled is found by multiplying the average velocity by the time for which the force acts.

20 (a) Explain what is meant by the *resultant of a system of forces*. State the parallelogram of forces law for the determination of the resultant of two non-parallel forces acting in the same plane. Describe an experiment to verify the law.

(b) A pulley system has a velocity ratio of four and an efficiency of 80 per cent. Calculate the value of the effort required to raise a load of 48 N.

(Joint Matriculation Board)

(a) See units 4.1 and 4.2.

(b) Use the equation for efficiency at the end of unit 5.7.

21 The following results were obtained in an experiment to investigate the performance of a single string pulley system (block-and-tackle).

Load N	50	150	250	350	450
Efficiency %	50	75	83.3	87.5	90

Plot a graph of efficiency against load.
Use your graph to determine
(a) the efficiency when the load is 200 N;
(b) the load for which the efficiency is 86%.

(Associated Examining Board)

(a) Draw a line through the 200 N point on the load axis parallel to the efficiency axis. Where this line cuts the graph draw a line parallel to the load axis and read off the value on the efficiency axis.

(b) A similar procedure to part (a) is followed, but in reverse.

22 A simple machine enables an effort of 40 N to lift a load of 300 N through a vertical height of 8.0 m in 32 s with uniform velocity.
Calculate:
(i) the mechanical advantage of the machine;
(ii) the power output of the machine.
The efficiency of the machine is 60%
Calculate:
(a) the power input to the machine;
(b) the distance through which the point of application of the effort moves in this time.

Describe in detail how you would carry out an experiment to determine the efficiency of any machine, with which you are familiar, for a particular load.

Suggest reasons why the efficiency of the machine you have chosen is less than 100%.

(Oxford and Cambridge)

(i) See unit 5.6.

(ii) See units 3.7 and 3.11. $\text{Power output} = \dfrac{\text{load} \times \text{distance load moves}}{\text{time}}$

(a) $\text{Efficiency} = \dfrac{\text{power output}}{\text{power input}} \times 100\%$

(b) $\text{power input} = \dfrac{\text{effort} \times \text{distance effort moves}}{\text{time}}$

The efficiency of machines is reduced by two factors: the weight of any moving parts and friction in the pivots.

23 An experiment was carried out to investigate the performance of a single-string pulley system (block-and-tackle) with a velocity ratio of five. The following results were obtained:

Load N	Effort required N
50	30
100	40
150	50
200	60
250	70

(a) For each value of the load calculate (i) the mechanical advantage; (ii) the efficiency. Give your results clearly in a table.

(b) Draw a graph of efficiency against load, using a vertical scale of 1 cm to represent 10% and a horizontal scale of 1 cm to represent 25 N.

(c) Showing your method clearly, use your graph to estimate
 (i) the efficiency obtained when raising a load of 175 N;
 (ii) the load for which the efficiency would be 55%.

(d) Assuming that the low efficiency was due only to the raising of the lower block during the operation of the system, deduce the weight of the lower block.

(e) What is the other principal cause of loss of efficiency of the pulley system?

(Associated Examining Board)

(a) See unit 5.6.

(b) ———————

(c) (i) Draw a vertical line through the x axis at the point corresponding to a load of 175 N. Draw a horizontal line through the point of intersection of the vertical line and the graph, and read off the value of efficiency on the y axis.
 (ii) Use a similar procedure to (i) but in reverse.

(d) As the pulley system has a velocity ratio of five it would have a mechanical advantage of five if 100% efficient. Thus the effort of 30 N would be sufficient to raise a load of 150 N etc. The difference between this and the actual load (i.e. 100 N) is the weight of the lower block.

(e) Friction in the bearings.

24 The diagram illustrates the lever employed in rowing a boat. Assuming that F marks the position of the fulcrum, mark the position of the load by L and that of the effort by E.

Assuming that the picture is drawn to scale, and making any necessary measurements, estimate the velocity ratio of the lever. V.R. = ...

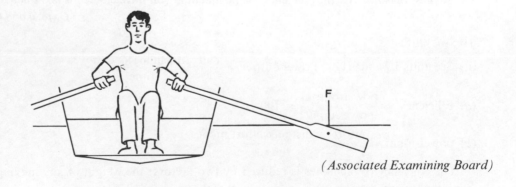

(Associated Examining Board)

The load acts at the rowlocks or crutches on the side of the boat and the effort at the rower's hand. The velocity ratio is normally about 1.5.

25 (a) (i) Describe carefully how you would find the centre of mass of an irregularly shaped lamina (sheet) of uniform thickness.

(ii)

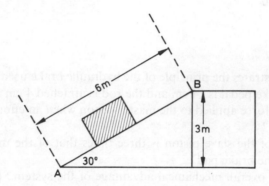

Fig.(i)   Fig.(ii)

The diagrams show a half-metre rule to which two large metal blocks are taped. In the situation of Fig. (i) it is very difficult to balance the rule, but it is quite easy in Fig. (ii). Explain carefully why this is so.

(b)

The diagram shows a packing case of weight 500 N on the inclined plane AB. Calculate how much work would be done in sliding the packing case up the inclined plane from A to B. The frictional force is 50 N.

State, giving your reason, whether the packing case would remain stationary if it were placed half-way up the inclined plane.

What effect would increasing the length of the inclined plane from B to the ground (keeping B at a height of 3 m above the ground) have on
 (i) the velocity ratio, and
(ii) the efficiency of the inclined plane considered as a machine?

Give a reason for each of your answers. (You may assume that the frictional force does not change.)   *(University of London)*
(a) (i) See unit 5.4.
    (ii) In Fig. (i) the centre of mass of the rule and blocks is vertically above the knife-edge. Any slight displacement of the rule results in a moment which tends to increase the displacement, hence this arrangement is unstable. In Fig. (ii) the centre of mass is vertically below the knife-edge and a slight displacement results in a moment which tends to decrease the displacement. This is a stable arrangement.
(b) The component of the weight acting down the plane is 500 sin 30°N. To this must be added the frictional force.
    The packing case would remain stationary half-way up the plane, if the frictional force were greater than the component of the weight acting along the plane. This is not the case as drawn.
  (i) The velocity ratio is the ratio of the length of the plane to the vertical height gained. It increases.
  (ii) The work output of this machine is constant as the packing case is lifted through a vertical distance of 3 m. However as the plane increases in length the work input increases. This is because the same frictional force has to be overcome for a greater distance. The efficiency thus decreases.

Unit 7

26 (*a*) (i) Describe how you would demonstrate experimentally that, for a body in equilibrium, the sum of the clockwise moments about a point is equal to the sum of the anticlockwise moments about the same point.

(ii) You are given a metre rule, a knife-edge and a known mass. How would you find the mass of the rule?

(*b*)

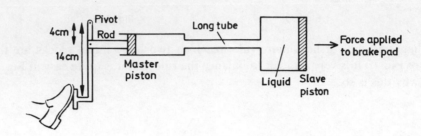

The diagram illustrates the principle of the hydraulic brake used in many cars. The effective length of the brake pedal is 14 cm and the rod is attached 4 cm from the pivot:

(i) What is the force applied to the master piston when an effort of 200 N is applied to the brake pedal?

(ii) If the area of the slave piston is three times that of the master piston, what force is applied to the brake pad?

(iii) What is the overall mechanical advantage of the system? (Assume that the system is 100% efficient.) *(University of London)*

(*a*) (i) See unit 5.1.

(ii) Balance the rule on the knife-edge to find the position of its centre of mass. Then place the knife-edge under a point between the centre of mass and one end of the rule. Move the known mass along the rule near the short end until balance is achieved. Apply the law of moments about the position of the knife-edge.

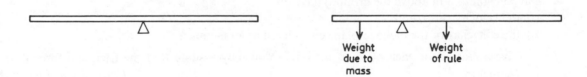

(*b*) (i) See unit 5.6. The velocity ratio is 3.5. Assuming 100% efficiency the force applied to the master piston is 700 N.

(ii) The liquid transmits the pressure on the master piston to the slave piston. Here it acts over three times the area, thus applying a force three times as great as at the master piston. Use the formula $\text{Pressure} = \dfrac{\text{Force}}{\text{Area}}$ for solution.

(iii) The force applied increases from 200 N at the pedal to 2100 N at the pad. The mechanical advantage is $\dfrac{2100}{200} = 10.5$.

27 Describe how you would set up and use a simple barometer to measure the atmospheric pressure.

(Joint Matriculation Board)

See unit 7.4.

Unit 8

28 (*a*) The diagram below, which is drawn full-scale, illustrates a wooden cube floating in water.

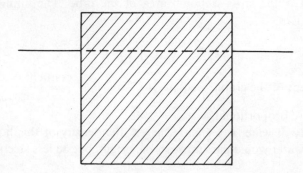

By making any necessary measurements of the drawing, calculate the relative density of the wood.

Relative density = ...

(Associated Examining Board)

(*a*) According to Archimedes' principle the upthrust equals the weight of water displaced. Thus the weight of the cube equals the weight of water displaced. The definition of relative density given in unit 6.2 can be written:

$$\text{relative density} = \frac{\text{weight of a body}}{\text{weight of an equal volume of water}}$$

or in this case, relative density $= \dfrac{\text{weight of water displaced}}{\text{weight of water with same volume as cube}}$

$$= \frac{\text{volume of water displaced}}{\text{volume of cube}}$$

$$= \frac{\text{depth of water displaced}}{\text{total height of cube}} = \frac{3}{4}$$

29 (*a*) A flat bottomed tube (area of base 2.5 cm²) has lead shot placed in it and is floated upright

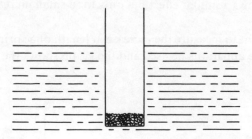

in a liquid whose density is greater than that of water. The following results are obtained:

Mass of empty tube   20 grams

Mass of shot in grams	10	13	16	19	22	25
Depth (*h*) immersed in centimetres	10	11	12	13	14	15

  (i) Plot a graph of the total mass of the tube and lead shot against *h*.

 (ii) Find the slope of the graph, and, either by using this or otherwise, find the density of the liquid.

(iii) What general conclusion concerning the variation of pressure with depth can you draw from your graph?

(iv) On the same axis sketch a graph showing what you would expect had the liquid been water.

(*b*) Describe a simple manometer and explain how you would use it to measure the excess pressure of the gas supply in the laboratory. You should state the liquid you would use.

(University of London)

(a) (i) ————
(ii) The weight of liquid displaced by the tube equals $hAdg$, where d is the density of the liquid and A the cross-sectional area of the tube. The application of Archimedes' principle gives

$$mg = hAdg$$
$$\text{or} \quad m = hAd$$

The gradient of the graph $\dfrac{m}{h}$ equals Ad. Thus $d = \dfrac{\text{gradient of the graph}}{\text{area } A}$

(iii) The two are proportional.
(iv) The density of water is 1 g/cm³ whereas the density of this liquid is 1.20 g/cm³. The graph for water would pass through the origin, but be less steep.

(b) See unit 7.3.

30 A block of metal (density 2700 kgm⁻³) has volume 0.04 m³. Calculate the mass of the block and its apparent weight when completely immersed in brine (density 1200 kgm⁻³).

(Southern Universities' Joint Board)

The mass of the block is the product of its volume and density.

The mass of liquid displaced is the product of the volume of the block and the density of brine. Its weight is the product of its mass and the acceleration due to gravity.

The upthrust equals the weight of liquid displaced. The apparent weight is the true weight less the upthrust.

Unit 9

31 Describe briefly how the length of a light spiral spring changes when the spring is stretched by a force which is gradually increased in magnitude. Would you expect the same comments to apply to a straight steel wire as it is stretched?   *(Southern Universities' Joint Board)*

See unit 9.1.

A straight wire shows a similar effect but only for a small increase in length.

32 John does an experiment to measure the increase in length of a spring as he loads it with masses. The diagram shows the apparatus he used and the graph shows the results he obtained.

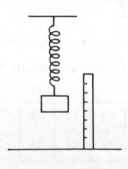
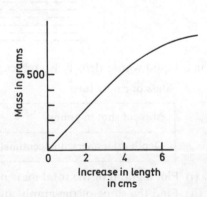

(a) Comment on the following statement. 'The apparatus could be used, together with the graph, to measure a downward force of about 5 N but might not be suitable for measuring greater forces.'
(b) Show, with the help of a labelled diagram, how you would use the apparatus to measure a horizontal force of about 3 N, such as the friction between a board and a bench.
(c) Given two springs of the kind that John tested, they could be arranged either in series or in parallel, as shown in the following diagrams.

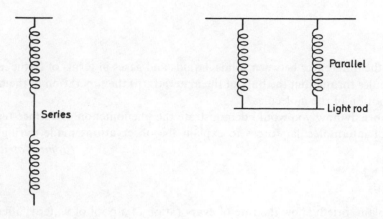

Series

Parallel

Light rod

   (i) You load each of these arrangements with masses. Sketch a graph with lines, showing the result you would expect for each of these two arrangements.
   (ii) Which arrangement would be more suitable for measuring a force of about 0.5 N? Give a reason.
   (iii) Which arrangement would be more suitable for measuring a force of about 9 N? Give a reason.
 (d) Suppose John's original single-spring apparatus is taken to the Moon. Would the apparatus be suitable for finding the mass of a sample of Moon-rock of about 1 kg? Give a reason.
 (e) Would the apparatus be suitable for measuring the force of attraction between two magnetised lumps of Moon-rock in contact? Assume the lumps to have a mass of 1 kg each and a force of attraction of about 1 N.   *(Oxford and Cambridge Nuffield)*

(a) The force due to gravity acting on a mass of 500 g is 5 N. The graph is straight up to this value showing that the spring is behaving elastically, and can be used to measure such a force. It would not be suitable for greater forces as these would permanently lengthen the spring.
(b) The spring is firmly fixed to the board at one end and the horizontal force applied at the other. A ruler clamped above the spring is used to measure its extension and the corresponding force found from the graph. Other arrangements can be used.
(c)   (i) The series arrangement is effectively a spring of twice the length. Thus twice the extension is obtained for corresponding forces compared to the single spring. The gradient is half that of John's graph.

        The parallel arrangement means that each spring experiences only half the total force applied. Thus half the extension is obtained for corresponding forces compared to the single spring. The gradient is twice that of John's graph.
      (ii) The series arrangement gives a greater extension per unit force and is thus more suitable for small forces.
      (iii) The parallel arrangement will give a straight line graph up to 10 N and is more suitable.
(d) Gravitational attraction on the surface of the Moon is about one-sixth of that on the Earth's surface. Thus on the Moon a mass of 1 kg experiences a force of about 1.7 N. John's apparatus is suitable.
(e) The total attraction due to gravity and magnetism is about 2.7 N. Thus the apparatus is suitable.

Unit 10

33

Aluminium

Camphor

 (a)   (i) The thin sheet of aluminium foil shown above will float on water in a bowl, and if a small piece of camphor is placed as shown the aluminium foil will move about on the surface of the water. Explain these observations.
      (ii) Explain why it is not sensible to rub the canvas of a tent in wet weather.
                                                                      (University of London)

See unit 10.

Unit 11

34 (*a*) Describe the differences between solids, liquids and gases in terms of (i) the arrangement of the molecules throughout the bulk of the material; (ii) the separation of the molecules; and (iii) the motion of the molecules.

(*b*) Describe briefly how you would demonstrate the phenomenon of surface tension. Use the concept of intermolecular forces to explain the observations made during your demonstration. *(Joint Matriculation Board)*

(*a*) See unit 11.2.

(*b*) See unit 10.

35 State and explain, briefly, how the rate of evaporation of a pool of water is affected by (*a*) the area of surface exposed to the atmosphere and (*b*) the humidity of the atmosphere.
(Southern Universities' Joint Board)

(*a*) If the pool has a large area of surface exposed to the atmosphere, then a higher proportion of its molecules are near the surface. Thus a higher proportion of the more energetic ones are able to escape from the surface. The rate of evaporation will be high.

(*b*) If the humidity of the atmosphere is high, it already contains a high concentration of water vapour and will be unable to take up much more. The rate of evaporation will thus be slow.

36 (*a*) Explain why, when a small quantity of oil is placed on the surface of water, the oil spreads to form a continuous film.

(*b*) (i) Explain briefly one reason for believing that the thinnest films are one molecule thick.

(ii) Describe fully a laboratory experiment which would enable you to estimate an upper limit for the diameter of an oil molecule.

(*c*) Give a short account of an experiment which demonstrates the Brownian motion of smoke particles in air. Explain how we can deduce from these observations that the molecules of a gas are in continual random motion. *(Oxford Local Examinations)*

(*a*) The force of attraction between water molecules and oil molecules is greater than the force of attraction between two oil molecules. The water molecules surrounding the oil pull outwards on the oil molecules with greater force than the other oil molecules exert. The oil is thus pulled out to make a larger surface.

(*b*) (i) If a very small drop of oil is placed on a clean water surface, it goes on spreading as long as the oil surface is continuous. The greatest area is when all the oil molecules form a single layer; that is the layer is one molecule thick.

(ii) See unit 11.1.

(*c*) See unit 11.3.

37 'The botanist, Robert Brown, while examining pollen particles floating on water, noticed through his microscope that they were in a continuous state of rapid, random motion.'

'When the stopper is removed from a bottle containing liquid ammonia, the ammonia, after a time, can be smelled in all parts of a room.'

(i) Show how these and similar observations lead to a 'kinetic' theory of the structure of liquids and gases.

(ii) Outline briefly the kinetic theory of gases and liquids.

(iii) Show how the theory accounts for the pressure exerted by a gas on the walls of its container.

(iv) Show how the theory accounts for the evaporation of a liquid.

(Oxford and Cambridge)

(i) See units 11.3 and 11.4. The smell must be due to some ammonia having moved across the room. This means liquid ammonia must consist of small particles (molecules) some of which have escaped. Also gaseous ammonia must consist of molecules which are moving. The fact that the smell takes some time to reach all parts of the room is due to the ammonia molecules continually colliding with air molecules on their way, which greatly reduces their average velocities.

(ii) See unit 11.1.
(iii) See unit 11.1.
(iv) See unit 11.1.

38 (*a*) Under a low-power microscope, smoke particles appear to move in a random manner. Use the kinetic model for a gas to explain this.

(*b*) A little bromine is released at the bottom of a tall cylinder containing air. Describe what happens and use the model to explain it.

(*c*) Use the model to explain how air enclosed in a motor car tyre exerts a pressure.

(*d*) The air in each of the four tyres of a motor car exerts a pressure above atmospheric of 3×10^5 N/m². Each tyre is in contact with the ground over an area of 80 cm² (0.008 m²). Estimate the mass of the car.

(*e*) What would happen to the area of contact between the tyres and the ground if the temperature rose? Explain your answer.   *(Oxford and Cambridge Nuffield)*

(*a*) See unit 11.3.

(*b*) See unit 11.4.

(*c*) The air molecules in motion inside a motor car tyre collide with the tyre walls. The change of momentum they undergo causes a force (and thus pressure) on the inside of the walls. The molecules in the atmosphere cause a similar effect outside. However, the density of the air inside is greater than that outside, and hence there are more collisions on the inside of the walls and thus a greater pressure.

(*d*)
$$\text{Pressure} = \frac{\text{force}}{\text{area}}$$

Thus the force between each tyre and the ground is the product of the pressure and area of contact with the ground. This force equals one quarter of the weight of the car.

(*e*) If the temperature rose all molecules would be moving faster making more frequent and more violent collisions with the walls. The pressure would rise and a smaller area of contact would support the weight of the car.

Unit 12

39 The diagram illustrates a thermostat used to control an electric immersion heater. Given the following thermal expansivities,

$$\begin{aligned}
\text{aluminium} &= 23 \times 10^{-6}/\text{K} \\
\text{copper} &= 17 \times 10^{-6}/\text{K} \\
\text{invar} &= 1 \times 10^{-6}/\text{K} \\
\text{steel} &= 12 \times 10^{-6}/\text{K}
\end{aligned}$$

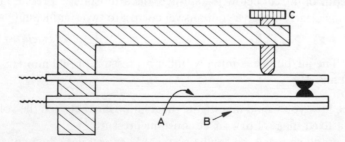

suggest a suitable metal for *A* and a suitable metal for *B*.

(*a*) Describe the effect of rotating screw *C*.

(*b*) What type of appliance could the thermostat shown in the diagram control if materials *A* and *B* were inter-changed?   *(Associated Examining Board)*

Use the two metals whose expansivity differs the most. *A* should be made from the metal with the higher value.

(*a*) See unit 12.1.

(*b*) A refrigerator.

Pivot

Steel rod

Mercury

40 (*a*) Describe how you would find by experiment the linear expansivity of a metal in the form of a rod or tube. State the precautions you would take and explain how the final value would be calculated.

(*b*)

The arrangement shown (right) is suggested as being suitable for use as a clock pendulum, the period of which is constant whatever the temperature. Suggest reasons why this might be possible.

Give and explain two physical reasons why water could not be used in place of mercury in such a pendulum. (*University of London*)

(*a*) See unit 12.2.

(*b*) Both steel and mercury expand as the temperature increases. The depth of mercury is adjusted so that the downward expansion of the rod is exactly compensated by the upward expansion of the mercury. The centre of mass of the pendulum remains at the same distance from the pivot.

Water is most dense at 4°C. Above and below this temperature it expands. If the temperature fell below 4°C the steel would be contracting and the water expanding. Even above 4°C the expansion of water is non-linear with temperature.

The density of water is much less than that of steel. A large volume of water would be required to have an effect similar to the mercury.

Unit 13

41 (*a*) A cylinder of internal volume 0.5 m^3 contains oxygen at a pressure of 70 times that of the atmosphere. What volume of oxygen at atmospheric pressure has been used when the pressure in the cylinder has fallen to 30 times that of the atmosphere, the temperature remaining constant? Explain your method of calculation clearly. (*Southern Universities' Joint Board*)

(*a*) See unit 13.1. Use Boyle's law, $PV = $ constant, for a fixed mass of gas.

The oxygen remaining in the cylinder would have to be compressed into 3/7 of the volume of the cylinder to restore it to a pressure of 70 times atmospheric pressure.

Thus 4/7 of the initial volume of gas at 70 times atmospheric pressure has been used. At atmospheric pressure this occupies 70 times this volume, or 40 times the volume of the cylinder; that is 20 m^3 (Boyle's law).

42 An air-bubble of volume 0.72 cm^3 is released at a depth of 14.0 cm below the surface of mercury in a vessel. Assuming there is no change of temperature calculate the volume of the air-bubble on reaching a depth of 4.0 cm below the surface, the atmospheric pressure being 760 mm Hg. Show your working clearly and give your answer correct to two significant figures.

(*Associated Examining Board*)

Use Boyle's law. The air-bubble is initially under a pressure of 900 mm Hg, and finally under a pressure of 800 mm Hg.

43 (*a*) (i) Describe how you would investigate the relationship between the volume and temperature of a fixed mass of dry air at constant pressure.
(ii) Draw a graph showing the results you would expect, labelling your axes.
(iii) Suppose you then used another pure gas (*e.g.* hydrogen or helium). What feature would the two graphs have in common?
(*b*) A cylinder of volume 0.4 litre contains a mixture of petrol vapour and air at atmospheric pressure and a temperature of 27°C. The gas is compressed suddenly by a piston until the volume is 0.05 litre and the pressure 20 times atmospheric pressure. What is the new temperature of the gas? (Assume the mixture obeys the gas equation.) (*University of London*)

(*a*) (i) See unit 13.2.
(ii) See unit 13.2.
(iii) Both graphs when extended would pass through − 273°C.

(b) Apply the universal gas equation – unit 13.4. $P_1 = 1$ atmosphere, $V_1 = 0.4$ litre, $T_1 = 300$ K, $P_2 = 20$ atmospheres, $V_2 = 0.05$ litre.

44 (a) With the aid of a labelled diagram, describe the structure and operation of a constant volume air thermometer.

(b) Name and state the law which defines the relationship between the volume and pressure of a fixed mass of gas at constant temperature.

(c) A certain mass of dry air has a volume of 1000 cm^3 when its temperature is 27°C and its pressure is 160 mm Hg below atmospheric pressure. Calculate its volume when its temperature is 87°C and its pressure is 40 mm Hg above atmospheric pressure.

(Associated Examining Board)

(a) See unit 13.3.

(b) Boyle's law. See unit 13.1.

(c) Apply $PV/T =$ constant – unit 13.4. Remember the temperature must be expressed in degrees Kelvin.

Unit 14

45 Explain the meaning of the term specific heat capacity of a substance.

Describe how you would determine the specific heat capacity of aluminium. You may, if you wish, take the specific heat capacity of water as known.

An electric kettle, having a 1.5 kW heating element, contains 925 g of water. With the heater on, the rate of rise of temperature of the water when its temperature is 60°C is 20.2°C per minute. After the heater is switched off the temperature falls and the rate of fall, again at 60°C, is 1.4°C per minute.

What would be the rate of rise of temperature at 60°C if there were no heat loss?

Calculate a value for the specific heat capacity of water, given that the heat capacity of the kettle including the heater is the same as that of 75 g of water.   *(Oxford and Cambridge)*

See the beginning of unit 14.

See the end of unit 14.1.

The kettle loses heat at 60°C at the same rate whether or not the heater is on. Thus the rate of rise of temperature would be 21.6°C per minute (20.2 + 1.4).

The effective mass of water is 1000 g (925 + 75). The specific heat capacity (c) is calculated using the equation

$$VIt = mc(\theta_2 - \theta_1), \quad \text{where } (\theta_2 - \theta_1)/t = 21.6/60$$

46 A 180-watt heater and a thermometer were immersed in 0.5 kg of water in a copper calorimeter. The following readings were obtained:

Temperature/°C	30	36	40	45	49	54	57
Time/Minutes		3	4	5	6	7	8	9

On the graph paper provided plot a graph of temperature against time. Start the axis from the origin and draw the best straight line through your points.

Using your graph, or otherwise, find

(a) room temperature (the temperature at which heating started), and

(b) the specific heat capacity of water.

Give two reasons why the value obtained for the specific heat capacity is more than the accepted value.

State two precautions you would take in carrying out this experiment to ensure a more accurate value for the specific heat capacity.   *(University of London)*

(a) Extend the graph back until it cuts the temperature axis at time $t = 0$.

Read off the temperature value.

(b) See unit 14.1. Use the equation $VIt = mc(\theta_2 - \theta_1)$ where $(\theta_2 - \theta_1)/t$ is the slope of the graph.

The rate of rise of temperature is smaller than it should be because
(i) some heat is given to the calorimeter, and
(ii) some heat is lost to the surrounding air.

The results therefore indicate that more heat has been supplied to the water than is actually the case.

Surround the calorimeter as far as is possible with good insulation. Stir the water before a temperature reading is taken to ensure the true temperature of the water is recorded.

Unit 15

47 (a) Explain
(i) why a kettle of water with a steady supply of heat takes a much longer time to boil dry than it does to reach its boiling point;
(ii) how evaporation differs from boiling; and
(iii) how the molecular theory of matter accounts for the drop in temperature which results when rapid evaporation of a volatile liquid occurs.

(b)

A lump of copper of mass 0.5 kg is placed in an oven for some time and then transferred quickly to a large dry block of ice at 0°C. When the temperature of the lump of copper

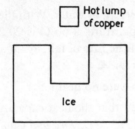
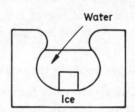

reaches 0°C, 0.3 kg of the ice is found to have melted. Estimate the temperature of the oven. (Take the specific heat capacity of copper as 400 J/kgK and the latent heat of fusion of ice as 3.2×10^5 J/kg.) Give two reasons why this estimate might be lower than the actual temperature of the oven.   *(University of London)*

(a)   (i) The latent heat of vaporisation of water is greater than the heat required to raise the temperature of 1 kg of water by about 80°C (80 × specific heat capacity). It thus takes longer for the kettle to boil dry than to reach its boiling point.
(ii) See unit 15.8.
(iii) See unit 15.8. When evaporation is rapid the liquid has no time to absorb heat from the surroundings. The molecules left have a lower average kinetic energy.
(b) The water formed remains at 0°C. The heat lost by the copper in cooling is all used to melt ice. Thus $m_c(\theta - 0)c_c = m_i L_i$ where θ is the oven temperature. The copper loses heat during transfer.

48 A compound originally solid, of mass 8 kg receives heat at a constant rate of 4.8×10^5 J/minute. Its temperature during the first 16 minutes of the experiment is shown in the graph.

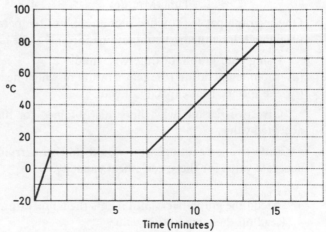

(*a*) Indicate clearly on the graph the portions of the graph which represent
  (i) the liquid state
  (ii) the solid state
  (iii) fusion or melting
  (iv) vaporisation or boiling
(*b*) Calculate those of the following quantities which can be determined from the graph and the information given above; if a property cannot be determined, simply state this as your answer.
  (i) the specific heat capacity of the substance
    in the solid state
    in the liquid state
    in the vapour state
  (ii) the melting point of the solid
  (iii) the boiling point of the liquid
  (iv) the specific latent heat of fusion or melting
  (v) the specific latent heat of vaporisation at boiling point *(Associated Examining Board)*

(*a*) Between minutes 1 and 7 heat is being absorbed but there is no change in temperature. Similarly after the 14th minute. Two changes of state thus take place (latent heat being absorbed). The first change of state must be from solid to liquid and the second, at a higher temperature, from liquid to vapour.
  (i) Between the seventh and fourteenth minutes the temperature is rising at a steady rate indicating the liquid state.
  (ii) In the first minute the temperature is rising at a steady rate indicating the solid state.
  (iii) Between the first and seventh minutes there is no increase in temperature, although energy is being supplied at a steady rate. This indicates fusion or melting.
  (iv) After the fourteenth minute there is no increase in temperature although energy is still being supplied at a steady rate. This indicates vaporisation or boiling.
(*b*) (i) See unit 14.1. The rate of increase in energy of the compound is equal to the rate at which heat is supplied. Use the equation, power $= \dfrac{mc(\theta_2 - \theta_1)}{t}$ where $\dfrac{\theta_2 - \theta_1}{t}$ is the slope of the relevant section of the graph.

The graph does not extend to the vapour state so this value cannot be determined.
(ii) and (iii) Read these from the horizontal sections of the graph.
(iv) and (v) Calculate the energy absorbed by the substance during the change of state by multiplying the constant rate of energy supply by the time taken. Then divide by the mass to determine the energy required to change each kilogram.

49 Define *specific latent heat* of steam and state a unit in which it may be expressed.

A pressure cooker consists of a container with a lid which has a hole in it. This hole may either be left open to the atmosphere or closed by a special plug.

When heat is supplied to the water in this cooker at the steady rate of 144 W, it is found that, with the hole closed, the temperature reaches a steady value of 100°C and the pressure inside is then atmospheric (760 mm Hg). In this steady state, what is happening to the heat energy being supplied?

Suppose now that the hole in the lid is opened and the rate of supply of heat is increased to 1500 W. At what rate will water boil away? Give your answer in grammes per minute.

(Specific latent heat of steam 2.26×10^6 J/kg)   *(Oxford and Cambridge)*

See unit 15.2.

After the temperature reaches a steady value of 100°C, heat is being lost to the surroundings at the same rate as it is being supplied.

After the rate of supply is increased the **extra** energy supplied each second is used to boil the water.

Thus the extra energy supplied each second $= \dfrac{mL}{t}$ where $\dfrac{m}{t}$ is the mass of water which boils every second.

50 Define *specific latent heat of vaporisation, saturated vapour, saturated vapour pressure* and *boiling point*.

Water at about 60°C is placed in a suitable enclosure, the air pressure in which can be varied. It is found that when the pressure has been reduced to a certain extent, the water begins to boil. Explain why this happens, where the latent heat of vaporisation comes from, and why the boiling eventually stops.

State one way in which you could start the water boiling again.

The early mountaineers used a portable apparatus for measuring the boiling point of water, so as to determine their height above sea level. Explain very briefly the reason why this method of height-determination is possible. *(Oxford Local Examinations)*

See units 15.1, 15.9 and 15.8 (+ 11.2).

A liquid boils when its saturated vapour pressure equals the surrounding air pressure. The value of the saturated vapour pressure of the water is determined by its temperature (60°C). As the surrounding air pressure is reduced it eventually falls to this value and the water begins to boil. Since no heat is being supplied from outside, the latent heat of vaporisation has to come from the water itself, and it therefore cools. The cooler water has a lower saturated vapour pressure, which no longer equals the surrounding air pressure and boiling stops.

Boiling would re-start if the air pressure were further reduced or the water heated.

Atmospheric pressure falls with increasing height above sea-level. Thus the boiling point of water is less the greater the altitude (unit 15.7).

51 (*a*) State two similarities and two differences between the processes of evaporation and boiling.

State two ways in which the rate of evaporation of a liquid could be increased.

(*b*) Describe experiments, one in each case, to show the effect on the boiling point of water of (i) dissolving common salt in the water, (ii) reducing the air pressure over the pure water. State the result you would expect in each case. *(Joint Matriculation Board)*

(*a*) See unit 15.8.

This can be increased by warming the liquid or passing a current of air over its surface.

(*b*) (i) See unit 15.5.
　　(ii) See unit 15.7.

Unit 16

52

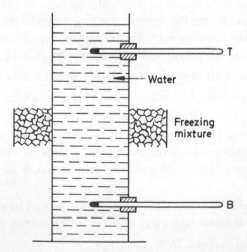

The diagram illustrates an apparatus which may be used to investigate the behaviour of water when slowly cooled. During such an experiment, the readings taken were as follows:

Reading	A	B	C	D	E	F	G	H	J
Time after start of experiment in mins.	0	10	20	30	40	50	60	70	80
Reading of top thermometer T in °C	10	10	9.8	9	8	6.6	4	0.4	0
Reading of bottom thermometer B in °C	10	7	4.6	4	4	4	4	4	3

(a) Plot the readings on a graph, temperature vertically using 1 cm to represent 1°C and time horizontally using 1 cm to represent 5 minutes. Draw and label clearly the two curves representing the two temperature readings.

(b) Explain very briefly what is happening:
   (i) between readings *A* and *C*;
   (ii) between readings *C* and *G*;
   (iii) between readings *G* and *H*;
   (iv) between readings *H* and *J*.

(c) What physical phenomenon is illustrated by this experiment?

(d) Why is this phenomenon important to fishes?   (*Associated Examining Board*)

(a) ——————

(b) See unit 12.3.
   (i) The water near the freezing mixture cools and its density increases. As a result it falls and is replaced by warmer water from below. The reading of the bottom thermometer falls.
   (ii) The water, from the freezing mixture downwards, is all at about 4°C by *D*, and convection in the lower half stops. The water in the upper half is now slowly cooled by conduction of heat from the top to the freezing mixture. The reading of the top thermometer falls.
   (iii) By *G* all the water is at 4°C. The water near the freezing mixture is cooled below 4°C, its density decreases and it rises. The reading of the upper thermometer continues to fall.
   (iv) By *H* all the water above the freezing mixture is near 0°C. The water in the lower half begins to cool by conduction. The reading of the lower thermometer begins to fall again.

(c) The unusual expansion of water. The density of water decreases as it cools from 4°C to 0°C.

(d) Water has its greatest density at 4°C. The water at the bottom of a pond is generally near this temperature even in the most cold weather. The bottom of a pond seldom freezes and fish are able to survive below the surface ice.

53 (a) (i) On a cold night in winter condensation occurs on the inside of the windows of a car. Explain carefully why this happens. Would you expect the effect to be greater if there were four people in the car instead of one? Give your reasons.

   (ii) House windows are often made of two sheets of glass with an air gap in between (double glazing). State, with reasons, what effect you would expect this to have on the condensation on the surface of the glass facing into the room.

   (iii) If it were possible to remove the air from between the two sheets of glass what difference would you expect? Give reasons.

   (iv) State and explain two other beneficial effects which double glazing has over a single sheet of glass.

(b) Why is frost more likely to occur on a clear night than on a cloudy night? Explain why, under these conditions, the frost is usually more severe in lower lying areas.

                                                           (*University of London*)

(a) (i) The air in contact with the windows is at a lower temperature than the rest of the air in the car. The cooler air has a lower saturated vapour pressure and is unable to contain all the water vapour in it; some condenses out. With four people in the car there will be more water vapour in the air than with one, so the effect will be greater.

   (ii) The air gap reduces conduction through the window, as air is a poor conductor. The inner pane of glass remains almost at room temperature, so less condensation occurs here.

   (iii) Even less conduction would occur due to the complete lack of a medium. Any convection currents in the gap would also be eliminated. Even less condensation would take place.

   (iv) It reduces heat losses and cuts down the noise entering the house.

(b) On a clear night the Earth loses a great deal of heat by radiation, whereas on a cloudy night this heat is reflected from the clouds. Cold air is more dense than warm air; thus the temperature in low lying areas is lower.

54 A concave reflector with radius of curvature 10 cm is placed in a ripple tank. Draw *scale diagrams* to illustrate the effect of each of the following:

(*a*) A series of plane wavefronts 2 cm apart are sent towards the reflector (see diagram).

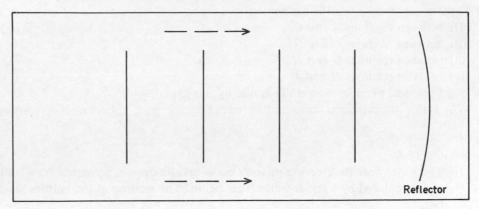

(*b*) A small spherical dipper placed 10 cm from the reflector on its axis sends out waves with wavelengths 2 cm.

(*c*) The dipper placed 15 cm from the reflector on its axis sends out waves of wavelength 2 cm.

(Oxford and Cambridge)

See unit 17.3. The focal length equals half the radius of curvature, *i.e.* 5 cm. In (*a*) the waves converge on the focal point after reflection. In (*b*) the dipper is at the centre of curvature of the reflector; thus the waves will be reflected normally and return to a point at the position of the dipper. In (*c*) the dipper is beyond the centre of curvature and so after reflection the waves converge on a point between the centre of curvature and the reflector.

In all cases the wavelength of the waves remains equal to 2 cm after reflection.

55 Fig. 1 represents a ripple tank in which the water is deeper in the region *A* than in the region *B* beyond the curved boundary.

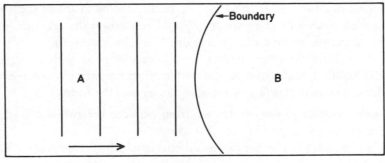

Fig. 1.

Plane waves are shown moving from left to right in *A*, where their speed is greater than it will be in *B*.

Draw a freehand diagram showing the shape of the waves which have passed into *B*.

(Oxford and Cambridge)

See unit 17.4. The waves slow down on entering the shallower region of *B*. The centre of each wave reaches *B* first and hence slows down first. In *B* the centre of each wave thus lags behind.

56 Explain what is meant by the *interference* of two wave motions, and by an interference pattern.

In order to show a stationary interference pattern between ripples on the surface of water, you are provided with the following apparatus: a ripple tank with a glass base, a vibrator which can make two 'dippers' move up and down, either exactly in step or exactly out of step with one another, a stroboscope (or strobo flash lamp) and a white screen. Describe how you would set up the apparatus, and draw to full scale an accurate diagram of the pattern you would expect to see if the two dippers are in step, are set 20 mm apart, and are producing ripples of wavelength 12 mm.

Explain briefly how you would expect the pattern to be altered if each of the following changes was made *alone*, with other conditions remaining unchanged:
(*a*) the stroboscope was flashing a little faster than required;
(*b*) the stroboscope was flashing a little slower than required;
(*c*) the dippers were oscillating at the same frequency but exactly half a period out of step;
(*d*) the dippers were oscillating in step with each other but at exactly half the original frequency.

(Oxford Local Examinations)

See unit 17.5. The diagram is made by drawing two sets of concentric circles whose radii differ by 12 mm and whose centres are 20 mm apart. The intersections of the sets of circles can be taken to represent reinforcement of the wave motions.

The stroboscope enables glimpses of the wave pattern to be seen at regular intervals. It is started slowly and its speed increased until a stationary picture of the wave pattern is seen.

(*a*) The glimpses come at slightly too frequent intervals. The waves do not quite move forward one position between glimpses and thus appear to be going slowly backwards.

(*b*) As (*a*) except in reverse. The waves appear to creep forward.

(*c*) The centre line of the pattern becomes a region of destructive interference – calm water; and all other regions interchange.

(*d*) The distance between wave crests is twice as great. The 'fan' effect of calm water and waves becomes more spread out and consequently fewer are seen.

57 (*a*)

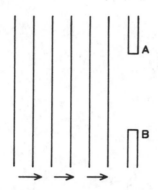

(i) Plane waves strike the aperture *AB*. Sketch the wave pattern emerging from *AB*. On the same diagram show by dotted lines what would be the effect of doubling the wavelength.

(ii) *CD* is now placed in the aperture, so that in effect *AC* and *DB* are two parallel slits. The wavelength is returned to its original value.

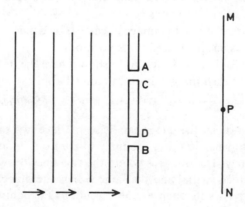

Draw the wave pattern in the region of the slits.

(iii) If a detector were moved along the line *MN* what would you expect to find? (Pay particular attention to the point *P* which is equidistant from *C* and *D*.) What differences would you expect along *MN* if the wavelength were shortened?

(*b*) Describe a demonstration to illustrate the difference between transverse and longitudinal waves. Give an example of each, other than those you have used in your demonstration.

(University of London)

(a) (i) See unit 17.6. Some diffraction occurs, but this increases when the wavelength is doubled, as the gap is now relatively narrower.

   (ii) A pattern is seen similar to the one produced by having a dipper in place of each of the two gaps (unit 17.5).

  (iii) If a detector were moved along *MN* alternate maximum and minimum wave amplitudes would be detected. A maximum would occur at *P* and the minimum points would be almost zero. Shortening the wavelength would lead to the maximum and minimum points being closer together.

(b) See unit 17.1.

58 In a ripple tank, plane waves meet a straight barrier in which there is a narrow gap which is much less than one wavelength wide. Draw a diagram showing the wave fronts in the water that is beyond the gap.   *(Oxford and Cambridge)*

If the gap is much less than one wavelength wide, it will act as if it were a point source of waves; that is the diffraction at the gap leads to semicircular waves beyond it.

Unit 18

59 On the diagram below, draw *two* rays from the point *A* to illustrate how the eye sees the image of the top of the candle.

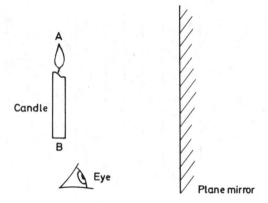

Also draw the image of the whole candle and mark *B′* the image of the point *B*.
    (Associated Examining Board)

SEE UNIT 18.2

60 An object is placed 15 cm in front of (i) a plane mirror; (ii) a concave mirror of radius 60 cm, and (iii) a convex mirror of radius 60 cm.

(a) Describe how you would locate the image in case (i) by experiment. State clearly the nature and relative size of this image, as well as its position.

(b) Draw scale diagrams for both (ii) and (iii) to show where the images are produced. State the nature and position of each image. State a practical application to each case.

    (Southern Universities' Joint Board)

(a) Look into the mirror so that the image can be seen. Place two pins in line with the direction of the image (use a drawing board or mounted pins). Look in at a different angle and place two more pins similarly. The image is where the two lines drawn through the pairs of pins meet. The image is erect, virtual and the same size as the object. It is on the perpendicular to the mirror which passes through the object and as far behind the mirror surface as the object is in front of it.

(b) See unit 18.3. (ii) Virtual, erect, magnified, and laterally inverted; (iii) Virtual, erect, diminished, and laterally inverted. Applications, (ii) shaving mirror; (iii) driving mirror.

Unit 19

61 A light ray is incident at various angles at the surface of a perspex block. The corresponding angles of incidence and refraction are listed below. Complete the table and then plot sin *i* against sin *r*.

Angle of incidence i	18°	37°	64°
$\sin i$	0.3090	0.6018	
$\sin r$	0.1994	0.3987	
Angle of refraction r	11°30′	23°30′	37°

Hence or otherwise calculate the refractive index of perspex.   *(Associated Examining Board)*

Look up and record the values of sin 64° and sin 37°. Plot sin i as the vertical axis and sin r as the horizontal axis. The graph will go through the origin and its slope is sin i/sin r, which is the refractive index of perspex (see unit 19.1).

62 Draw a diagram showing the refraction of plane light waves at a plane boundary such as that between air and glass.

Use the diagram to establish a relationship between the refractive index of glass and the speeds of light in air and in glass. Explain why Snell's law sets a limit on the direction of waves in glass which can emerge into the air, *i.e.* determines the critical angle. What is the value of the critical angle for a glass of refractive index 1.60?   *(Oxford and Cambridge)*

See unit 19.1. The relationship is established in unit 19.1. Critical angle is discussed in unit 19.2. Sin $C = 1.0/1.6 = 0.625$.

63 (*a*) The refractive index of glass is 1.5. What do you understand by refractive index?

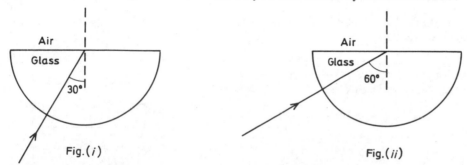

Fig.(*i*)   Fig.(*ii*)

In the above diagrams a ray of light incident along a radius undergoes no change in direction on entering the glass. However a change does take place on entry. What is this?

When each of the rays travelling in the glass strikes the plane face there is a change in direction. Determine the directions of the rays shown when they emerge from the glass, and show these on sketch diagrams of Figs (*i*) and (*ii*). Find also the angle of incidence on the plane face for which the ray would emerge along the face of the block.

(*b*) Draw ray diagrams to show:
   (i) the appearance of a straight stick partially immersed in water at an angle less than 90° to the surface;
   (ii) how a right-angled isosceles glass prism can be used to turn a ray of light through 90°. What are the advantages of using a prism rather than a silvered mirror for this purpose?
                                                    (University of London)

(*a*) See unit 19.1. On entry the velocity of the light decreases. In case (*i*) refraction takes place according to the usual equation. In case (*ii*) total internal reflection takes place and the ray leaves the glass through the semicircular face in a direction at 60° to the normal shown.

When the ray emerges along the face of the block $r = 90°$, whence sin $i = 1/1.5$.

(*b*) (i) See unit 19.1.
   (ii) The ray is incident at 90° to one of the faces making up the right angle. It is undeviated at this face but strikes the next one at 45°, which is greater than the critical angle; hence it is internally reflected through 90°. It strikes the third face normally and passes through it undeviated.

A silvered mirror causes multiple images due to reflection at the front and back of the layer of glass. A prism forms one clear reflected image.

64 (*a*) Describe an experiment to establish the relationship between the angles of incidence and refraction for light travelling from air to glass. State the result you would expect to obtain.

 (*b*) (i) Draw a ray diagram to show the action of a simple magnifying glass.

 (ii) A converging (convex) lens has a focal length of 0.25 m (25 cm) and is held close to the eye like a magnifying glass. Find how far from the lens an object must be placed so that the image is seen at a distance of 0.25 m (25 cm).   (*Joint Matriculation Board*)

(*a*) See unit 19.1.

(*b*) (i) See unit 19.6.

 (ii) Either construct a ray diagram as described in unit 19.5 or use the formula $\dfrac{1}{u}+\dfrac{1}{v}=\dfrac{1}{f}$

from the same unit. As the image is virtual its distance carries a negative sign.

65 Explain the meaning of the terms *principal focus* and *focal length* as applied to a converging lens.

Describe how you would determine the focal length of a converging lens. Illustrate your description with a ray diagram.

A camera makes use of a lens of focal length 100 mm. How far from the film must the lens be placed for it to produce sharp images of very distant objects?

How far, and in which direction, must the lens be moved from this position for it to produce a sharp image of an object 1.10 m from the lens?   (*Oxford and Cambridge*)

See unit 19.3.

See unit 19.4.

For distant objects the film must be placed in the plane of the focal point; that is 100 mm behind the lens.

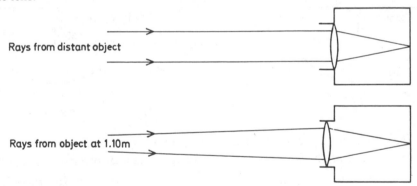

Rays from distant object

Rays from object at 1.10m

As the object comes closer, so the film must be moved back. Use the formula in unit 19.5 to calculate v and subtract 0.1 m (100 mm) from the value obtained.

66 (*a*) Describe two methods by which you could determine the focal length of a converging (convex) lens, the first a quick but approximate method, the second a more accurate method.

 (*b*) An object 3.0 cm tall is placed 10.0 cm from a diverging (concave) lens of focal length 6.0 cm so that it is perpendicular to, and has one end on, the principal axis of the lens. *Either* by a ray drawing on graph paper *or* by calculation using a clearly stated sign convention, determine the height, position and character of the image.

(*Associated Examining Board*)

(*a*) See unit 19.4.

(*b*) See unit 19.5, for either method of determining the image distance. If calculating the image distance, the focal length of the concave lens carries a negative sign. Using the real is positive sign convention, if the image distance is negative, the image is virtual. A virtual image will be erect.

$$\frac{\text{height of image}}{\text{height of object}}=\frac{\text{image distance}}{\text{object distance}} \qquad \text{(unit 19.5)}$$

67 (a) Describe fully, showing the arrangement used, an accurate method of finding the focal length of a converging (convex) lens.

(b) Explain what is meant by a virtual image. Show by means of a ray diagram how a lens can produce such an image. State a use for the arrangement you describe.

(c) Explain what is meant by a long-sighted eye and show how a suitable lens can correct this defect.
(University of London)

(a) See unit 19.4.

(b) Sometimes, after refraction through a lens, rays from a point diverge rather than converge. The point these rays appear to come from after refraction is the position of the virtual image. A virtual image may not be projected on to a screen, but can only be seen by an eye looking into the lens. Magnifying glass.

(c) Long sight occurs when the image formed by the eye falls behind the retina. A weak converging lens bends the rays in a little before they enter the eye.

68 An object 3.0 cm tall is placed 7 cm from a converging (convex) lens of focal length 4.0 cm. *Either* by means of an accurate scale ray drawing on graph paper *or* by calculation using a clearly stated sign convention, determine the position, height and character of the image.
(Associated Examining Board)

See unit 19.5.

69 Describe *one* way you would demonstrate experimentally that light travels in straight lines.

A pinhole of 1 mm diameter is in the middle of a piece of black paper covering one end of a tube 1 m long. The other end of the tube is covered by a screen of tracing paper. When this 'pinhole camera' is pointed towards the sun, the diameter of the image on the screen is found to be 10 mm. Draw a ray diagram (not to scale) showing how the image is formed.

What would you expect the image diameter to be if the length of the tube were increased to 2 m?

What would be the diameter of the sun's image formed by a converging lens of focal length 2 m? Mention *two* ways in which the image formed by the lens differs from that in the pinhole camera.
(Oxford and Cambridge)

See unit 18.1.

The pinhole camera is described in unit 18.1. If the depth of such a camera is doubled, the diameter of the sun's image will double. The lens will form an image with the same diameter as in the case of the 2 m tube. The lens image will be brighter and will be slightly coloured due to the lens bending different colours through slightly different angles.

70 (a) In each of the parts of the figure, drawn to the same scale, rays of light are shown striking a lens. Copy the drawings and show in each case possible directions of the rays when they come out of the lens.

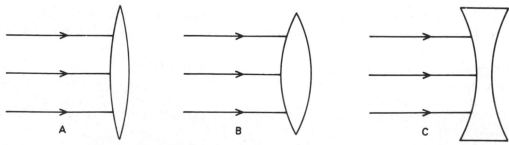

(b) Lens *D* in the figure below forms an image *I* of a very small bright object *O*.

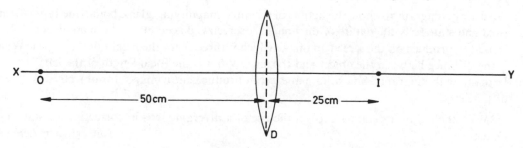

(i) If the small object is raised 2 cm above the line *XY*, where is the new image position?

(ii) Draw a sketch to show how the lens forms the image of the object in its new position. You should show the path of two rays from the object.

(iii) If the object is moved a short distance further away from the lens, what happens to the image?

(c) Draw a suitable diagram to show how such a lens can be used as a magnifying glass. The rays in your diagram should show the direction in which the light waves travel.

(d) What evidence leads us to believe that light is a wave motion?

(e) On the basis of the wave theory, explain how light is converged by a lens.

(Oxford and Cambridge Nuffield)

(a) Lenses *A* and *B* converge the rays to a point on the axis; *B* converges more than *A*. *C* diverges the rays so that they appear to have come from a point on the axis to the left of *C*.

(b) (i) The ray from the small object which passes through the centre of the lens will be undeviated. By geometry the new image position is 1 cm below *I*.

(ii) Draw two rays from the new position of the object, which are deviated by the lens to both pass through the new position of the image.

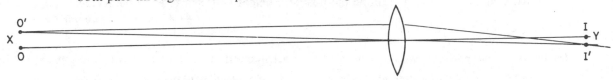

(iii) The image will move closer to the lens. It also becomes slightly smaller.

(c) See unit 19.6.

(d) Interference and diffraction -- see units 20.1 and 20.2.

(e) The velocity of light is less in glass than air. Those parts of the wavefronts, which pass through the converging lens near its centre, travel more slowly for a greater distance than those parts of the wavefronts which pass through the edge of the lens. The result is that the rays (which are normal to the wavefronts) converge.

71 (a) Draw a ray diagram to show how a plane mirror forms an image of a point source of light. On your diagram show clearly the position of the image and explain what is meant by saying the image is virtual.

How does the mirror placed adjacent to the scale and beneath the pointer of some instruments assist in the taking of accurate readings?

(b) A projector is used to produce an image 0.96 m by 0.72 m of a slide 40.0 mm by 30.0 mm. The screen is 3.0 m from the projection lens. Determine the distance between the slide and the projection lens and the focal length of the lens.

What features in the design of a projector ensure that a well-illuminated image is produced on the screen? *(Oxford and Cambridge)*

(a) See unit 18.2. If one ensures that the image of the pointer in the mirror is exactly behind the pointer, it means that viewing is normal, that is at 90° to the scale. There is thus no error due to parallax.

(b) A projector is basically a single lens system. The linear magnification required is 24. This means that $v/u = 24$ (by geometry). As $v = 3.0$ m, $u = 12.5$ cm. The lens formula (unit 19.5) is used to calculate f.

Two condenser lenses are used to concentrate the light from the lamp onto the slide (unit 19.7).

72 (a) Give a ray diagram to show the action of a simple magnifying glass. Name the type of lens used and state fully the nature of the image. A screen is placed 80 cm from an object. A lens is used to produce on the screen an image which is three times the height of the object. What is the distance between the object and the lens? What is the focal length of the lens?

(b) Explain, with diagrams, how a human eye can produce clear images for objects at different distances.

With the aid of a diagram explain the use of a diverging lens in correcting a named eye defect. *(University of London)*

(a) See unit 19.6. A short focal length converging lens is suitable. The image is erect, virtual and magnified.

$u + v = 80$ cm and magnification $= v/u = 3$. u and v are found using these two equations. f is found using the lens formula (unit 19.5).

(b) See unit 19.9. Short sight.

73 Describe, with the help of a labelled diagram in each case, (a) the optical system of the human eye and (b) a simple form of photographic camera.

Explain how the eye accommodates in order to see clearly objects over a range of distances, and how the camera is adjusted so that objects at different distances can be photographed.

A camera with a simple lens of focal length 50 mm is set so as to give a sharp picture of an object 1 metre away. How far is the film from the lens? In which direction (closer to the film, or further from the film) would the lens have to be moved next, in order to photograph a very distant object?   (*Oxford Local Examinations*)

(a) See unit 19.9.
(b) See unit 19.8.
Use the lens formula (unit 19.5) to find u. Towards the film.

74 (a) Explain the meaning of the term *focal length* as applied to a convex (converging) lens and a concave (diverging) lens.

Describe briefly an experiment you would use to measure accurately the focal length of a convex (converging) lens.

(b) Draw a ray diagram to show how two lenses may be used to construct a compound microscope. State the type of each lens and its approximate focal length. Show on the diagram the positions of the principal foci of the lenses in relation to the object and image positions.

Specify the nature of each image.   (*Joint Matriculation Board*)

(a) See units 19.3 and 19.4.
(b) See unit 19.10. The object lens has a very short focal length (about 1 cm). The eye lens has a focal length of about 5 cm. The image formed by the object lens is real; that formed by the eye lens is virtual.

Unit 20

75 (a) (i) Parallel, straight waves are produced in a ripple tank, as shown in the figure. A triangular plate of glass (*ABC*) is placed in the tank so that the water over the glass is very shallow. Copy the diagram and complete the waves in the region *ABC*.

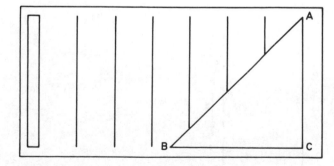

(ii) What happens to the wavelength as the waves cross the boundary?
(iii) What happens to the frequency as the waves cross the boundary?
(iv) If the wavelength is 2 cm and the frequency 10 Hz, what is the velocity of the waves?
(b) (i) When a beam of blue light falls on a prism the beam is bent as shown in the figure below. Explain this in terms of waves.

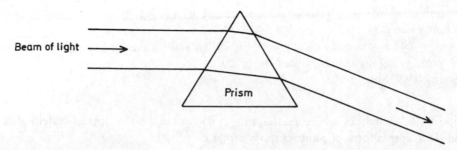

Beam of light →

Prism

(ii) If red light is used instead of blue, the red light is bent less than the blue light. Explain this.

(c)

Light passes through a slit and a lens, and then through a blue filter and a diffraction grating. Sharp images of the slit are seen on the distant screen as shown in the figure below.

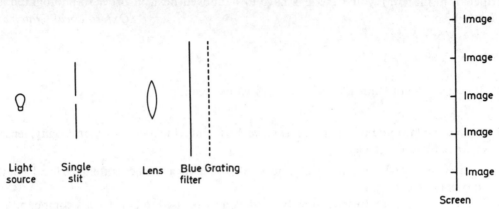

Light source Single slit Lens Blue filter Grating

Image
Image
Image
Image
Image
Screen

   (i) What is the purpose of the lens?

  (ii) Explain in terms of waves why there are several images.

 (iii) If red light is used instead of blue, the images are further apart. Explain this difference.

(Oxford and Cambridge Nuffield)

(a) (i) See unit 17.4.

  (ii) It decreases.

 (iii) It remains the same.

 (iv) Use the equation $v = f\lambda$, given in unit 17.1.

(b) See unit 19.1. (i) The lower part of the beam enters the prism first and thus slows down first. The beam thus bends downwards. The lower part leaves the prism last and is last to speed up. The beam thus bends downwards again. (ii) Red light does not slow quite so much as blue and is thus bent less.

(c) (i) The lens focuses the incident light to form an image of the single slit in the plane of the screen. (ii) See unit 20.3. (iii) Red light has a greater wavelength than blue. The angle θ must thus be greater before a path difference of one wavelength occurs.

76

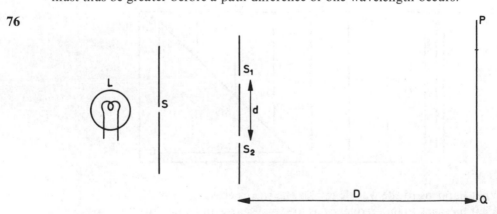

L

S

S₁

d

S₂

D

P

Q

The figure, not drawn to scale, represents an experiment for showing Young's fringes.

 (i) What would be a likely size for the distance d between the slits S_1 and S_2? What would be a likely size for the distance D from the slits S_1S_2 to the screen PQ?

(ii) Draw a labelled diagram to show what you would see on the screen *PQ*.

(iii) Mention three difficulties you might encounter doing this experiment.

(iv) If the lamp *L* were replaced by one giving light of a single colour, what could you find out about the light from the experiment?

(v) What do you think would happen if the slit S_2 were covered over leaving *S* and S_1 unchanged?

(vi) Two students look at yellow light. One says that yellow light is really made up of two kinds of light, red and green, added together. The other does not agree and says it is only one kind. Try to design an experiment which would decide between the two ideas. Make it clear how your experiment would enable you to decide.   (*Oxford and Cambridge Nuffield*)

(i) and (ii) See unit 20.2.

(iii) The alignment of *L*, *S* and the pair of slits S_1 and S_2.

Ruling S_1 and S_2 sufficiently parallel and close.

Avoiding stray light flooding the experiment.

(iv) The wavelength.

(v) There are now no interference fringes, as there is only one source.

(vi) Pass the yellow light through a pair of such slits, a diffraction grating or a prism and see if there is any sign of red and green light on the far side.

77 State *four* ways light, heat radiation and radio waves are similar and *two* ways in which those three types of radiation differ.   (*Associated Examining Board*)

Similar: They are transverse, show interference effects, travel with the same velocity, and do not need a medium through which to travel.

Dissimilar: They have different wavelengths and frequencies, and penetrate different distances through materials due to differing absorption.

78 The diagram below illustrates, very approximately to scale, the electromagnetic spectrum.

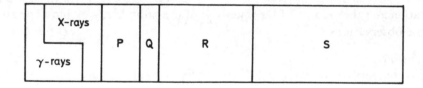

Name the parts of the spectrum which are marked only with letters.

Moving across the spectrum from left to right (γ-rays to *S*) what property of the radiation is (*a*) decreasing, (*b*) increasing, (*c*) constant?   (*Associated Examining Board*)

See unit 20.4.

Unit 21

79 (*a*) Explain what is meant by (i) a transverse wave; (ii) a longitudinal wave. Describe how a wave of each kind can be produced in a spiral spring.

(*b*) A tuning fork is set vibrating by striking the prongs on a cork. Describe the resulting vibrations (i) in the fork, (ii) in the surrounding air.

(*a*) See unit 17.1.

(*b*) See the beginning of unit 21.

80 (*a*) What is a progressive wave? With the aid of diagrams, explain the difference between *longitudinal* and *transverse* waves. Give *one* example of each type of wave.

(*b*)

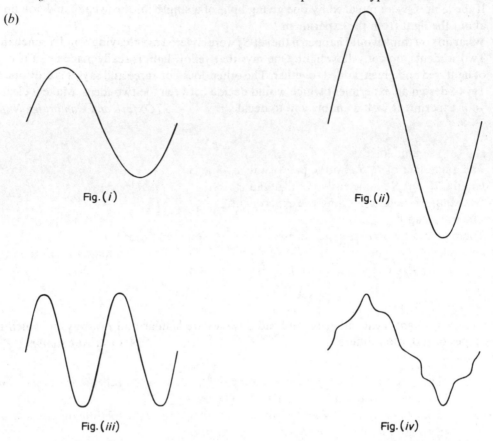

Fig. (*i*)   Fig. (*ii*)

Fig. (*iii*)   Fig. (*iv*)

The sound produced by musical instruments is picked up by a microphone and the waveform displayed on the screen of a cathode ray oscilloscope. The Figs. (*i*), (*ii*) and (*iii*) show the waveforms obtained when tones from the same instrument are displayed on the screen and Fig. (*iv*) the waveform from a different instrument. Compare the sound represented in Fig. (*i*) with the sounds represented by Figs. (*ii*), (*iii*) and (*iv*), taken in turn. State, giving reasons for your answer, the ways in which the sounds are similar and the ways in which they differ. (The *Y* amplification and the velocity of the time-base of the cathode ray oscilloscope are not altered during the observations.)   (*Oxford and Cambridge*)

(*a*) See unit 17.1.
(*b*) Discuss the relative amplitudes and frequencies of the four waveforms and whether or not they are pure notes.

81 How would you attempt to test in the laboratory the truth of the statements
   (*a*) that sound requires a material medium to transmit it;
   (*b*) that the speed of sound in air is independent of frequency?

Draw a graph to illustrate the changes of pressure with time at a fixed distance from a loudspeaker emitting a steady pure tone of 200 Hz.

Draw a second graph to show the effect of increasing the intensity, keeping the frequency constant.

Show on a third graph what happens if the tone is changed to one an octave higher in pitch but having the same amplitude as the original tone.   (*Oxford and Cambridge*)

(*a*) See the beginning of unit 21.
(*b*) Use a resonance tube (unit 21.8) with tuning forks of different frequency. Calculate the speed in each case. The graph will be a sine wave whose wavelength is about 1.5 m. Increasing the intensity increases the amplitude (or height) of the waves. The third graph will show a sine curve having half the wavelength (twice the frequency) of the first two and the same height as the first one.

82 In a cathedral, a man sitting 165 m from the organ and 5 m from the nearest loudspeaker notices a time delay of 0.5 seconds between the start of a chord transmitted by the amplification system and that transmitted through the air in the cathedral. Calculate the velocity of sound in the air in the cathedral and state what assumption must be made to enable this to be done.

(Associated Examining Board)

The time delay is due almost entirely to the extra 160 m the direct sound has to travel through air compared to that which comes through the amplification system. The velocity is obtained by dividing this distance by the time taken. The extra time taken by the sound coming through the amplification system is ignored as the velocity of the electrical pulses in the cable is very high and hence the time taken very small.

83 State *three* factors which affect the frequency of the note produced by a vibrating stretched string.

(Associated Examining Board)

See unit 21.6.

84 (*a*) With reference to a thin wire under tension which is vibrating between two fixed clamps, explain the terms: *node, transverse wave, stationary wave, amplitude, frequency.*

How, if at all, are the pitch and loudness of the note emitted by the wire affected if (i) the amplitude is increased, and (ii) the vibrating length of the wire is decreased?

(*b*) Calculate the speed of radio waves from the data for 'Radio Four': 1500 m; 200 kHz.

(Southern Universities' Joint Board)

(*a*) See unit 21.5 and the beginning of unit 21.
(i) Loudness increased, pitch unaltered; (ii) Frequency increased, the amplitude and hence loudness is likely to decrease.
(*b*) Use the wave equation at the beginning of unit 21.

85 (*a*)

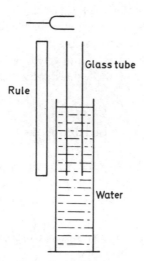

A tuning fork is struck and held over the glass tube shown above. The glass tube is then moved vertically, with the fork always held close, until the note heard becomes very loud. Name and explain this effect, and describe fully how it may be used to determine the speed of sound in air.

(*b*) Explain why
(i) strings of different thickness are used on a stringed instrument such as a violin;
(ii) notes of the same pitch played on a violin and a flute sound different;
(iii) the lowest pitched note played on a flute is lower than that played on a piccolo (a similar but smaller wind instrument). *(University of London)*

(*a*) See unit 21.8. The shortest length at which the note heard becomes very loud corresponds to the fundamental mode of vibration. In this case $l = \lambda/4$. The wave equation at the beginning of unit 21 is then used to determine the speed of sound from the frequency and wavelength.

(b) (i) Amongst other factors the frequency of the note from a violin depends on the mass of unit length of the string used (unit 21.6).

(ii) See unit 21.4.

(iii) In the case of wind instruments the air vibrates with fundamental frequency when $l = \lambda/4$, where l is the length of the vibrating air column for a particular note.

Hence $\lambda = 4l$ and the maximum wavelength obtainable from a flute is greater than from a piccolo, because the flute is longer. The lowest pitched note obtainable from a flute is thus lower than from a piccolo.

86 (a) Air vibrating in a narrow pipe, which is open at both ends, gives a fundamental note of frequency 256 Hz. Draw a diagram to illustrate the mode of vibration of the fundamental, another to illustrate that of the first overtone. Mark nodes and anti-nodes of displacement N and A respectively.

Calculate the frequency of the first overtone.

The length of the pipe is 66 cm. Calculate the velocity of sound in air, giving your answer in m/s correct to three significant figures.

(b) While the tension of a vibrating string was kept constant, its length was varied in order to tune it to a series of tuning forks. The necessary lengths are given below.

Tuning fork	C	D	E	G	C
Frequency of tuning fork (Hz)	256	288	320	384	512
Length of string (cm)	117	104	94	78	59

By the appropriate use of the above readings (graphical or otherwise), determine the relationship between the frequency of vibration and the length of the string.

(Associated Examining Board)

(a) The pipe is open at both ends and thus for the fundamental note $l = \lambda/2$ (unit 21.8) and for the first overtone $l = \lambda$. The wavelength of the first overtone is half that of the fundamental note. The first overtone thus has twice the frequency of the fundamental note.

Use the wave equation $v = f\lambda$ to calculate v.

(b) Comparing the values for the notes C and C' shows that when the length is halved the frequency doubles. Inspection of the values shows that frequency × length is constant (approx.). Thus $f \propto 1/l$.

Unit 22

87 Why does an iron ship become magnetised during the building process?

(Oxford and Cambridge)

During the construction of an iron ship much hammering and riveting takes place. As a result of this rough treatment the iron tends to become magnetised in the direction of the Earth's magnetic field, as the domains align themselves in this direction.

Unit 23

88 Describe, with the aid of a diagram, a form of electroscope, *e.g.* a gold leaf electroscope, and explain as fully as you can how it works.

How would you charge an electroscope positively by induction?

A continuous stream of α particles enters the earthed metal case of a charged gold leaf electroscope through a small hole in the side. Explain what happens to the leaf.

What would be the effect of covering the hole by successive layers of very thin paper?

(Oxford and Cambridge)

See unit 23.1.

The α particles will ionise the air near to the leaf. The charge on the leaf will attract ions of opposite sign. These will move to the leaf and steadily neutralise the charge on it, causing the leaf to fall. The ions similar in sign to the charge on the leaf will be repelled from it.

α particles are heavily ionising and thus short range. They will pass through a few layers of very thin paper but will be stopped as more are added.

The leaf would no longer fall.

89 (*a*) Describe how the potential over the surface of a charged conductor may be investigated, showing clearly the apparatus used.

What would your findings be for a positively charge pear-shaped conductor on an insulating stand?

(*b*)

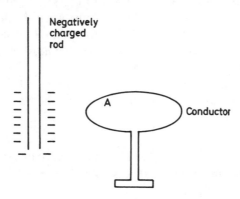

Negatively charged rod

A

Conductor

Draw a diagram to show any charge(s) on the above conductor when the charged rod is placed as shown.

The conductor is momentarily earthed at *A*. Draw another diagram to show any charge(s) now on the conductor. (The rod remains in position.)

(*c*) Describe fully, giving a circuit diagram, how an electric current may be used to magnetise strongly a steel bar. The polarity produced must be linked clearly with the current direction.

(University of London)

(*a*) See unit 23.2.

(*b*) The presence of the negatively charged rod results in the end of the conductor nearest it becoming positively charged and the opposite end negatively. When the conductor is momentarily earthed the negative charges on it are repelled to earth. The positive charge on the conductor is distributed with most of it on the end near the rod. This process is referred to as charging by induction.

(*c*) See unit 22.1.

90 (*a*)

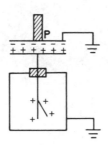

An earthed metal plate *P* is placed directly above the plate of a charged leaf electroscope.

Describe and explain what happens when

(i) *P* is slid slowly sideways;

(ii) *P* remains fixed in its original position, but a slab of insulator (*e.g.* paraffin wax) is slid slowly between the plates.

Describe briefly, preferably with the aid of a diagram, a practical capacitor which depends upon *one or both* of the above effects.

(*b*) Explain why

(i) the north seeking pole of a magnetised knitting needle, when suspended so that it can rotate in a *horizontal* plane, does not usually point to the geographical north pole, and

(ii) the same needle, when suspended so that it can rotate in a *vertical* plane, usually comes to rest at an angle to the horizontal. Describe what steps you could take to demagnetise the needle completely.

(University of London)

(*a*) See units 23.7 and 23.8. An electroscope measures potential. In this arrangement it is measuring the potential difference between its metal plate and the earthed plate above. The two plates may be considered as a parallel plate capacitor.

(i) As *P* is slid sideways the overlapping area of the plates is reduced, thus decreasing the capacitance of the arrangement. The same quantity of charge is being stored in a smaller capacitor; the potential must be greater (unit 23.7). The leaf rises.

    Alternatively, as *P* is slid sideways work is done in separating the charged plates. This work increases the potential energy of the system and the leaf rises.

(ii) The insulator lowers the potential difference between the plates (the capacitance of the parallel plate capacitor is increased). The leaf falls. See Fig. 23.10.

(*b*) (i) The knitting needle points to the earth's magnetic north pole which is several hundred miles from its geographical north pole.

(ii) The north and south magnetic poles of the earth are well below the surface. The earth's magnetic field is not usually therefore parallel to the earth's surface (Fig. 22.4(*a*)). See unit 22.2.

Unit 24

91 The charge on an electron is 1.6×10^{-19} C. How many electrons strike the screen of a cathode ray tube each second when the beam current is 16 mA?   *(Oxford and Cambridge)*

$$\text{Current} = \frac{\text{charge}}{\text{time}}; \text{ 16 mA means } 16 \times 10^{-3} \text{ C/s.}$$

Thus the number of electrons $= \dfrac{16 \times 10^{-3}}{1.6 \times 10^{-19}} = 10^{17}$ every second.

92 The figure represents a circuit containing four resistors and a 6 V battery of negligible internal resistance, connected as shown.

(*a*) Write down

    (i) the resistance of the path *ABC*;

    (ii) the current through *ABC*;

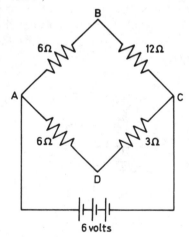

    (iii) the resistance of the path *ADC*;

    (iv) the current through *ADC*;

    (v) the p.d. between *A* and *B*;

    (vi) the p.d. between *A* and *D*;

    (vii) the p.d. between *B* and *D*;

(*b*) In which of the four resistors is there the greatest rate of conversion of energy into heat? Explain.   *(Oxford Local Examinations)*

(*a*) (i) Resistance in series (unit 24.5).

    (ii) Apply $V = IR$ (unit 24.3) to the branch *ABC*.

    (iii) Similar to (i).

    (iv) Similar to (ii).

    (v) The p.d. between *A* and *B* is found by applying $V = IR$ to the 6 Ω resistor, using the answer to (ii).

    (vi) Similar to (v) using the answer to (iv).

    (vii) The difference between the answers to (v) and (vi).

(b) Power = voltage × current. Multiply the current through each resistor by the p.d. between its ends using answers to (a) (ii), (iv), (v) and (vi).

or Twice the current flows via *D* as via *B*. As the potential difference between *A* and *C* is the same by either route, the rate of energy conversion in the branch *ADC* is twice that in the branch *ABC*. Two-thirds of the energy conversion in the branch *ADC* occurs in the 6 Ω resistor (using power = $I^2 R$). Thus the greatest rate of energy conversion occurs in the resistor between *A* and *D*.

93 Explain the meaning of the terms *electromotive force* of a cell, *internal resistance* of a cell.

Would you expect two identical cells in parallel to drive more current through a resistor than one cell does?

Why do two identical cells in series drive more current through a resistor than one does, and why do they not double the current?

A cell of 6.0 V e.m.f. and negligible internal resistance is connected to a resistor and drives a current of 3.0 A through it. Another cell of e.m.f. 1.5 V is inserted in the circuit in series with the first one. The current remains at 3.0 A. What is the internal resistance of the second cell?

(Oxford and Cambridge)

See units 24.2 and 24.5.

The p.d. across the resistor is the same with two cells in parallel as with one. The current through the resistor is thus the same in each case. Each cell supplies about half the current.

Two identical cells in series provide twice the e.m.f. that one cell does. The current through the resistor does not double because the circuit now includes the internal resistance of two cells rather than one.

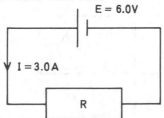 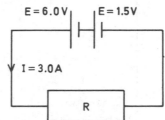

(a) Using $R = V/I$ gives $R = 2\,\Omega$.

(b) With both cells $E = I(R + r)$ where $E = 7.5$ volts, $I = 3.0$ A and r is the internal resistance of the 1.5 V cell.

94 (a) Draw a labelled circuit diagram to show how the power consumption of an electric heating coil (immersion heater) marked 12 V d.c., 30 W may be determined. Describe the procedure of the experiment and indicate the ranges of the meters you would use.

(b) The total resistance of the circuit shown in the diagram, measured between *P* and *Q* is 2 ohms. The resistances of *A* and *B* are 4 ohms and 6 ohms respectively.

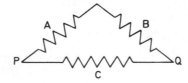

What is the resistance of *C*?

The total current flowing between *P* and *Q* is 2 A. What is the current in resistor *A*?

(a) See Fig. 24.3 where *R* represents the position of the immersion heater. The immersion heater should be placed in water to prevent it overheating and the readings of the ammeter and voltmeter taken.

It is shown in unit 24.8 that current = power/potential difference.

Thus $I = \frac{30}{12} = 2.5$ A. A voltmeter reading to 20 V and an ammeter reading to 5 A would be suitable.

(*b*) Apply unit 24.6 to find the value of *C*.

Use the equation $V = IR$. Between *P* and *Q* the total current is 2 A, and the total resistance 2 ohms. Thus the potential difference *V* is 4 V. Apply the equation $V = IR$ to the branch *A* + *B* to determine the current through *A*.

95 (*a*) Two heating coils dissipate heat at the rate of 40 W and 60 W respectively when connected in parallel to a 12 V d.c. supply of negligible internal resistance. Calculate the resistances of the coils.

Assuming that these resistances remain constant, what would be their rates of dissipation of heat when connected together in series with the same supply as before?

(University of London, part)

(*a*) See unit 24.8. Apply current = power/potential difference in each case and then apply $V/I = R$.

The resistances add up when placed in series. Calculate the current passing through them using $I = V/R$. Then use power $= I^2 R$ for each one.

Unit 25

96 On the diagrams below draw the magnetic field patterns for current flowing in (*a*) a straight

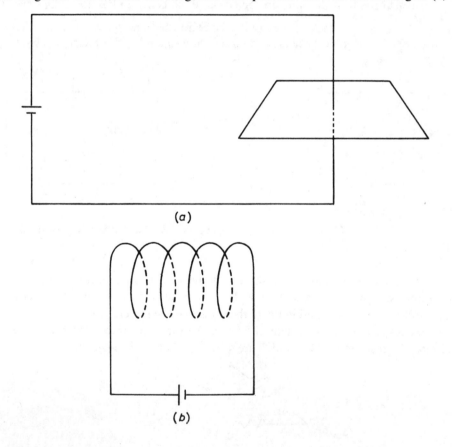

(*a*)

(*b*)

vertical wire, (*b*) a coil. On each diagram use arrows to indicate the direction of the field.

(Associated Examining Board)

(*a*) See the beginning of unit 25, also figures 25.1 and 25.3.
(*b*) See the beginning of unit 25.

97 (*a*) Draw a labelled circuit diagram of the vibrator of an electric bell or buzzer. Explain how the vibrator works.
(*b*) Describe a method of magnetisation not involving a permanent magnet. Use the domain theory of ferromagnetism to explain why the specimen has become magnetised.

(Joint Matriculation Board)

(*a*) See unit 25.2 and Fig. 25.5.

(*b*) See units 22.1 and 25.1 and Fig. 22.2. The domain theory suggests that clusters (domains), each of many millions of atoms, form small magnets within the specimen. When the specimen is not magnetised these domains have their magnetic axes lying in closed chains, so that there is no resultant magnetism. When the specimen is magnetised the magnetic axes are aligned, along the length of the specimen, to form open chains. As a result there are free poles at each end of the specimen.

98 Describe an experiment which shows that a wire carrying a current in a magnetic field may experience a force. Draw a diagram to show the direction of this force when the direction of the current is perpendicular to the field.

Explain how forces of this nature acting on a rectangular coil in a magnetic field can cause it to rotate.

Draw a labelled diagram of a d.c. motor, and explain clearly the features which produce continuous rotation of the armature coil.   *(Oxford and Cambridge)*

See Figs. 25.8 and 25.9 in unit 25.5.

The way in which forces of this nature cause a rectangular coil to rotate when it is in a magnetic field is explained in unit 25.6. Details of the d.c. motor are given in the same unit.

99 Describe with the help of a diagram the construction of a moving-coil galvanometer or ammeter. Point out *two* features which affect the size of the deflection for a given value of the current through the coil.

A galvanometer of resistance 20 ohms gives its full scale deflection for a current of 1 mA.

Explain how you would adapt it for use as a voltmeter to read up to 100 volts.

Has this voltmeter any advantage over one reading up to 100 volts on a similar scale and having a resistance of 1000 ohms? If so, what is it?   *(Oxford and Cambridge)*

See units 25.6 and 25.7. The strength of both the magnetic field and the coiled spring or torsion fibre determine the size of deflection.

See unit 25.8. The current through the meter must never exceed 1 mA. To achieve this when a potential difference of 100 volts is applied, a large resistance must be placed in series with the meter. The total resistance R, including that of the meter, is calculated using $R = V/I = 100/10^{-3}$. Thus $R = 10^5$ ohms. As the meter has a resistance of 20 ohms, the extra resistance added must be 99,980 ohms.

Yes. A voltmeter should have as large a resistance as possible to keep as small as possible the current it takes from the component whose potential difference is being measured.

100 In the diagram below, G represents a galvanometer whose resistance is 5.0 ohms and which

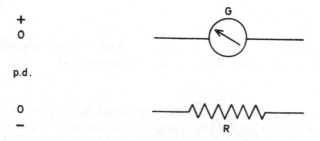

gives full scale deflection with a current of 10.0 mA. Calculate the resistance of the single resistor R which is necessary to enable the galvanometer to be used as a voltmeter reading up to 2.0 volts. Indicate on the diagram how the resistor R, the galvanometer and the potential difference to be measured, should be connected together.   *(Associated Examining Board)*

See unit 25.8. The method is similar to that used in the previous question.

101 (*a*) Describe fully the structure and mode of action of an ammeter which can measure *alternating* currents.

(*b*) A moving coil meter, for measuring direct current, has a full scale deflection 15 mA and a resistance of 5 Ω.

Explain how the meter could be converted
(i) to act as an ammeter reading up to 5 A;
(ii) to act as a voltmeter reading up to 5 V. *(University of London)*

(*a*) See unit 25.3.
(*b*) (i) The current through the meter must never exceed 15 mA. If it is to act as an ammeter reading up to 5 A, the remainder of the current (4.985 A) must pass through a resistor in parallel. See unit 25.8.
(ii) See unit 25.8. The method is similar to that used in the two previous questions.

Unit 26

102 (*a*) Draw and label a diagram of a simple d.c. generator (dynamo). Draw a sketch graph of the voltage output during two revolutions of the coil. What modifications would be necessary to make the generator produce a.c. instead of d.c.?

Draw a sketch graph of the a.c. voltage output which would then be produced during two revolutions of the coil.

(*b*) State Ohm's Law in words.

An electric heater is labelled 240 V 960 W. Assuming that the heater is in normal use, calculate (i) the current carried by the heater.
(ii) the electrical resistance of the heater.

If fuses rated 2 A, 5 A, 13 A, and 15 A, were available, which would be the best choice of fuse for the heater? *(Associated Examining Board)*

(*a*) See units 26.2 and 25.6; also Fig. 26.3.
See unit 26.3 and Fig. 26.3.
(*b*) See unit 24.3.
(i) Use the relationship, power = potential difference × current, to calculate the current (unit 24.8).
(ii) Use the relationsip $R = V/I$ (unit 24.3).
The fuse rating should exceed the expected current, but by as little as possible. A suitable value would be 5 A.

103 (*a*) State with reasons the material which you would consider most suitable for
(i) a permanent magnet,
(ii) the core of a transformer.
(*b*) What energy losses in a transformer are minimised by
(i) the use of low-resistance wire,
(ii) lamination of the core?
(*c*) Draw and clearly label a diagram of an electroscope.
(*d*) Describe with the aid of a series of simple diagrams how you would charge an electroscope by the method of induction, using a negatively-charged rod.

(Associated Examining Board)

(*a*) (i) See the beginning of unit 22. Some materials, such as steel, retain their magnetism once they have been magnetised. These are suitable as permanent magnets.
(ii) Other materials, such as soft iron and mumetal do not retain their magnetism and are suitable for the core of a transformer, where it is necessary for the magnetic field to reverse many times each second. It is essential that there is little or no residual magnetism in this case.
(*b*) See unit 26.5.
(*c*) See Fig. 23.1 in unit 23.1.
(*d*) See unit 23.1.

104 (*a*) Describe an experiment by which you could show how the heat produced by a steady current in a heating coil depends on the size of the current used.

(*b*)

240V ~ d.c.   1000 turns   50 turns   (V)   Lamp R = 4.0Ω   K

The circuit shows a step-down transformer used to light a lamp of resistance 4.0 Ω under operating conditions.

Calculate

(i) the reading of the voltmeter, *V*, with *K* open,

(ii) the current in the secondary winding with *K* closed (the effective resistance of the secondary winding is 2.0 Ω),

(iii) the power dissipated in the lamp,

(iv) the power taken from the supply if the primary current is 150 mA, and

(v) the efficiency of the transformer. *(University of London)*

(*a*) The basis of the experiment is described at the end of unit 14.1. The work of unit 24 is also relevant. A d.c. supply is connected to the heater. The voltage applied is varied thus varying the current. The temperature rises caused by different currents flowing for the same length of time are recorded. As the mass and specific heat capacity of the body being heated are the same throughout the experiment, the temperature rise is a measure of the heat produced.

(*b*) (i) $\dfrac{\text{Secondary voltage}}{\text{Primary voltage}} = \dfrac{\text{number of secondary turns}}{\text{number of primary turns}}$. This is a step down transformer with a turns ratio of 20:1. The voltage is stepped down by the same factor.

(ii) Apply $I = V/R$ where $R = 6.0$ ohms.

(iii) Power $= I^2 R$, where $R = 4.0$ ohms.

(iv) The power = primary current × primary voltage.

(v) Efficiency $= \dfrac{\text{power output (secondary)}}{\text{power input (primary)}} = \dfrac{\text{(iii)}}{\text{(iv)}}$.

105 (*a*) Draw a labelled diagram of a simple d.c. generator, showing clearly how the current is conducted to and from the generator.

Explain how the generator works, and name three factors which control the magnitude of the e.m.f. What changes would you make in the design to obtain a smoother output e.m.f.?

(*b*) Electricity is transmitted over long distances by using high voltage a.c. Explain the advantages of using high voltages and give reasons why d.c. is less suitable. *(University of London)*

(*a*) See unit 26.2. The strength of the magnets, and the number of turns and speed of rotation of the coil, all control the magnitude of the e.m.f.

(*b*) See unit 26.6. A transformer will not operate if a steady voltage (d.c.) is used (see answer to previous question). Thus the voltage cannot be stepped up or down.

Unit 27

106 How would you show that the particles emitted from the heated cathode of a thermionic diode are negatively charged? *(Oxford and Cambridge)*

Connect a power supply to the heater (normally 6.3 V d.c. or a.c.). A suitable potential difference is connected between the cathode and anode such that the anode is positive (Fig. 27.1*a*). A current flows. However if, by reversing the connections of this power supply, the anode is made negative with respect to the cathode, no current flows.

If the heater is off no current flows in either case. Thus, when the heater is on, negatively charged particles must be leaving the cathode; these are attracted by the anode when it is positive, but repelled when it is negative.

107 Make a diagram to show the construction of a thermionic diode and explain why current can pass through it in one direction only. Draw a diagram of the circuit you would set up to enable you to investigate the current–voltage relation for a diode.

Draw a graph in ink, plotting I vertically and V horizontally, to indicate the kind of result you would expect. Add a curve in pencil to show the effect of an increase in cathode temperature.

(Oxford and Cambridge)

See unit 27.2.

A suitable circuit is shown in Fig. 27.1*a* and a typical graph given in Fig. 27.1*b*. If the cathode temperature is increased more electrons flow, and at each voltage the current is greater.

108

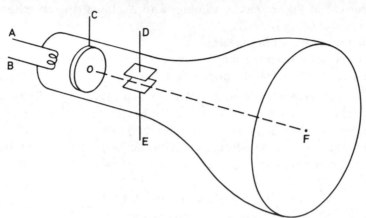

A manufacturer makes an electron 'deflection tube' which consists of an evacuated glass bulb with some electrodes in it. He states that the tube works best if a 6 V supply is connected between A and B and a 3 kV supply between A and C. The electrode at C has a small hole in it, and just beyond the hole there are two parallel metal plates with a connection to each of them.
(*a*) What is the purpose of having a potential difference between A and B?
(*b*) Which terminal of the 3 kV supply should be connected to C?
(*c*) Explain the purpose of the high voltage.
(*d*) Why is the tube 'evacuated'?

When the manufacturer's instructions are followed, a spot of green light appears at F on a special coating on the inside of this deflection tube.
(*e*) When a p.d. of 100 V is applied between D and E (D positive) the spot moves away from F. Draw a diagram to show the direction in which the spot moves.
(*f*) Explain why the spot moves in (*e*).
(*g*) How could you make the spot move further from F?
(*h*) If the spot is not very bright, what *two* separate changes might you make to increase the brightness?
(*i*) An alternating voltage of frequency 50 Hz is applied across the plates D and E. Describe what you would now see on the screen.
(*j*) The tube shown in the diagram could *not* be used to display an alternating voltage as a wavy line, as an oscilloscope does. What could be done to achieve this result?

(Oxford and Cambridge Nuffield)

See unit 27.3.
(*a*) The potential difference between A and B causes the filament to heat so that electrons are released.
(*b*) and (*c*) The positive terminal is connected to C so that the negatively charged electrons are accelerated towards it.
(*d*) The tube is evacuated so that electrons do not collide with gas molecules and lose energy.
(*e*) and (*f*) The negative electrons are attracted towards the positive plate D; hence the beam is deflected upwards.

(*g*) A bigger p.d. between *D* and *E* would cause the spot to deflect more. The same effect is achieved by decreasing the voltage applied to *C* so that the electrons pass between the plates more slowly.

(*h*) Either increase the heater current or the voltage applied to *C*. The first change produces more electrons, the second gives each electron more energy.

(*i*) A vertical line with its centre at *F*.

(*j*) It is necessary to provide a pair of vertical plates. A time base is provided by connecting a changing voltage to these plates thus causing the beam to swing from side to side.

109 What is the purpose of a time base of a cathode-ray oscilloscope? *(Oxford and Cambridge)*

See unit 27.5.

The time base repeatedly moves the beam across the screen at a particular speed. It thus provides a time axis, and enables one to see how a quantity (*e.g.* voltage) varies as time passes.

110 State briefly how you would use a cathode-ray oscilloscope to display: (i) the alternating voltage applied to a circuit; (ii) the alternating current through a circuit. *(Oxford and Cambridge)*

See unit 27.5.

(i) The alternating voltage is connected directly to the input to the *Y* plates and a suitable time base selected.

(ii) The input to the *Y* plates is connected to the opposite ends of a resistor in the circuit; that is the cathode-ray oscilloscope is in parallel with the resistor. The potential difference across the resistor appears on the screen, and is proportional to the current through the circuit.

Unit 28

111 (*a*)

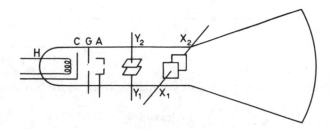

The diagram represents a cathode ray tube. Name the following parts and state and explain their functions:

(i) *H*   (ii) *G*   (iii) *A*.

A time base with a period of $1/25$ s is connected across the plates X_1 and X_2, and an alternating voltage of frequency 50 Hz is connected across Y_1 and Y_2. Draw a diagram of the trace obtained on the screen. Indicate on the diagram the quantities represented by the horizontal and the vertical movement of the spot.

(*b*) A radioactive source is thought to emit alpha and gamma radiation. Describe briefly simple experiments which you would perform to verify this. *(University of London)*

(*a*) See unit 27.5.

(i) The heater which emits electrons when it is hot.

(ii) The grid which controls the number of electrons reaching the anode and thus the brightness of the display.

(iii) The anode voltage controls the kinetic energy with which each electron passes through the tube beyond the anode *A*.

50 Hz means that 50 complete waves are produced every second. In the $1/25$ s it takes the spot to cross the screen, 2 waves are produced and seen on the screen. The vertical axis represents voltage, the horizontal, time.

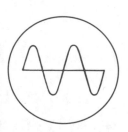

(*b*) See unit 28.4.

112 Complete the following table regarding the properties of atomic particles.

Particle	Mass	Charge	Location in atom
Electron			Orbit
Neutron	1		
		+1	Nucleus

(Associated Examining Board)

See section 28.2.

The electron has a very small mass (0 atomic mass units) and unit negative charge. The neutron has no charge and is located in the nucleus. The other particle located in the nucleus is the proton which has a mass similar to the neutron (1 atomic mass unit).

113 Carbon 14 is an isotope of carbon. Explain the meaning of this statement.

(Oxford and Cambridge)

See unit 28.3.

114 Complete the following table of the properties of alpha, beta and gamma emissions.

	Wave or particle	Sign of charge (if any)	Relative mass (if any)
Alpha			
Beta			
Gamma			

(Associated Examining Board)

See unit 28.4.

115 The diagram illustrates α, β and γ radiation emanating from a lead box containing radium, the

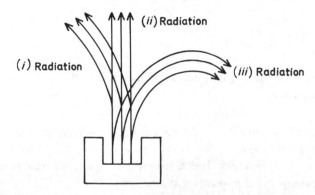

radiation being under the influence of a magnetic field perpendicular to the paper. (The results were not all obtained in one experiment.) Label the three types of radiation.

(Associated Examining Board)

γ radiation is part of the electromagnetic spectrum. It is uncharged and thus is not deflected by a magnetic field. It is represented by (ii).

(i) is likely to be α radiation and (iii) β radiation. They have opposite charges, and are thus deflected in opposite directions by a magnetic field.

116 Of the three kinds of radiation, α-particles, β-particles and γ-rays, emitted by a radioactive substance, state which one
(*a*) travels at the greatest speed;
(*b*) carries negative charge;

(c) forms dense straight tracks in a cloud chamber;
(d) is not deflected by a magnetic field;
(e) must be emitted when $^{232}_{90}$TH decays to $^{228}_{88}$Ra;
(f) is similar in nature to X-rays;
(g) is similar in nature to cathode rays   *(Oxford Local Examinations)*

(a) γ-rays being part of the electromagnetic spectrum travel with the speed of light.
(b) β-particles.
(c) α-particles – they are heavily ionising.
(d) γ-rays are part of the electromagnetic spectrum and as such are uncharged.
(e) A mass drop of 4 units and a charge drop of 2 units is due to the emission of an α-particle.
(f) γ-rays and X-rays are both parts of the electromagnetic spectrum.
(g) Cathode-rays are electrons, as are β-particles.

117 An atomic nucleus A consists of N neutrons and Z protons. How many of each will there be
   (a) after the nucleus has emitted an α-particle, (b) after the new nucleus has then emitted a
   β-particle?   *(Oxford and Cambridge)*

   (a) An α-particle is a helium nucleus, and as such contains two protons and two neutrons. The
   new nucleus thus contains $N - 2$ neutrons, and $Z - 2$ protons.
   (b) A nucleus emits a β-particle as a result of a neutron changing to a proton. The nucleus has
   one less neutron and one more proton as a result, that is $N - 3$ neutrons, and $Z - 1$ protons.

118 $^{12}_{6}$C and $^{14}_{6}$C represent two .... of the element carbon. $^{14}_{6}$C has the same number of
   .... as $^{12}_{6}$C but has two more .... $^{14}_{6}$C is radioactive, disintegrating with the emission

   of a β-particle to form $^{x}_{y}$N.
   Write out the above passage, inserting the missing words and the numbers x and y.
                                                           (Oxford and Cambridge)
   See unit 28.3.

119 (a) Describe the structure of a diode valve and state the conditions necessary for a current to
   flow between its electrodes. Draw a diagram of a circuit which could be used to test your
   statement.
       Explain briefly the use of a diode to rectify a.c. Include a graph showing how the output
   current varies with time.
   (b) Explain the significance of the numbers in the symbol $^{60}_{28}$Ni.
       Comment briefly on the fact that it is also possible to have $^{58}_{28}$Ni.
       Part of the uranium decay series is written $^{234}_{90}$Th → $^{234}_{91}$Pa → $^{234}_{92}$U → $^{230}_{90}$Th. State
   what particles are emitted at each stage of this decay.   *(University of London)*

   (a) See unit 27.2 particularly Figs. 27.1 and 27.2.
   (b) See units 28.5 and 28.3.

       An increase of one in the value of the atomic number indicates a β decay, a decrease of
   two indicates an α decay.

120 Give an account of the nature of the alpha, beta and gamma radiations emitted from radioactive
   materials.
       If a sample were believed to be radioactive, how would you confirm whether this was so
   and attempt to establish which forms of radiation were being emitted?
       A detector shows that the activity of a radioactive sample falls from 120 units to 15 units in
   12 minutes. Determine the half-life of the sample, explaining your calculation.
                                                           (Oxford and Cambridge)
   See unit 28.4.

       15 units are obtained by halving 120 three times, *i.e.* $15 = 120 \times (\frac{1}{2})^3$. Thus three half-lives
   elapse in 12 minutes. The half-life of the sample is 4 minutes.

121 (*a*) With the aid of a labelled diagram, describe the structure of a single-coil direct current motor including a split-ring commutator. Indicate on your diagram the directions of the conventional flow of current, the magnetic-field and the rotation produced.

If such a motor is disconnected from the d.c. supply and rotated at a steady rate, it acts as a dynamo. Illustrate by a sketch graph the variation with time of the output e.m.f. (the wave-form).

(*b*) The half-life of a certain radioactive substance is 4 years. Of an initial mass of 40 g of the substance, how much will remain unchanged after

(i) 4 years?

(ii) 8 years?

(iii) 16 years?

Plot a graph of the mass remaining unchanged against time. Hence estimate as accurately as possible the mass remaining unchanged after 14 years.

(Associated Examining Board)

(*a*) See unit 25.6.

See unit 26.2.

(*b*) (i) 20 g after one half-life.

(ii) 10 g after two half-lives.

(iii) 2.5 g after four half-lives.

The graph will be exponential in shape. Approximately 3.6 g remain after 14 years.

CSE questions

Unit 2

122 (*a*) Define the terms *velocity* and *acceleration*. Choose one of these quantities and explain what is meant when it is said to be 'uniform'.

(*b*) A car travels at a constant velocity of 20 m/s for 200 seconds and then accelerates uniformly to a velocity of 30 m/s over a period of 20 seconds. This velocity is maintained for 200 seconds before the car is brought to rest with uniform deceleration in 30 seconds.

Draw a velocity–time graph for the journey described above.

From the graph find

(i) the acceleration while the velocity increases from 20 m/s to 30 m/s,

(ii) the total distance travelled by the car,

(iii) the average speed over the entire journey.

(*a*) See units 2.4, 2.5 and 2.6.

(*b*) The graph is a horizontal straight line for the first 200 seconds, corresponding to a constant velocity of 20 m/s. It then rises steadily to 30 m/s after 220 s. The next section is a horizontal straight line for 200 s; the final section being a straight line falling from 30 m/s to rest in 30 s.

(i) The acceleration is the slope of the graph from 200 s to 220 s.

(ii) The total distance travelled by the car is found by calculating the area under the entire graph. This can be done either by estimating the number of squares under the graph or by calculating the areas of convenient simple shapes under the graph and adding them up.

(iii) Divide the total distance travelled (from ii) by the total time taken (450 s).

Unit 3

123 Explain the fact that a body on the moon has the same mass as, but would register less on a spring balance than, the same body on earth. Explain why the spring balance shows this difference and why a lever arm balance would not.

Mass is a measure of the quantity and type of material in a body, and neither of these are affected by the body's position in the Universe. The spring balance stretches in proportion to the force of attraction of the earth or moon for the body (weight). These forces are different.

A lever arm balance uses the principle of moments to compare the weight of the body with that of another body. Both bodies are attracted by either the earth or the moon, and no difference shows.

124 An electric railway locomotive of mass 50 000 kg starts from rest and after 20 seconds has accelerated uniformly to a speed of 25 m/s.

(a) Calculate
    (i) the acceleration of the locomotive;
    (ii) the horizontal driving force;
    (iii) the distance travelled in 20 seconds;
    (iv) the work done by the driving force in 20 seconds.

(b) When the locomotive reaches 25 m/s the brakes are applied until the vehicle stops.
    (i) How much energy is lost by the locomotive?
    (ii) What happens to this energy?

(c) Assuming the electrical energy was produced by a coal burning power station, briefly describe the energy changes which have taken place by the time the locomotive is travelling at 25 m/s.

(East Anglian Examinations Board (North))

(a) (i) $\text{Acceleration} = \dfrac{\text{change in velocity}}{\text{time taken}}$

    (ii) Force = mass × acceleration.
    (iii) The distance travelled is the average speed (12.5 m/s) multiplied by the time (20 s).
    (iv) Work done is force × distance moved against the force, (ii) × (iii). Alternatively the work done by the locomotive results in an increase in its kinetic energy ($\frac{1}{2}mv^2$).

(b) (i) It loses the kinetic energy it had acquired in the first 20 s; that is the answer to (a) (iv).
    (ii) It is converted to heat in the brake shoes. Some is dissipated by air resistance and friction in the carriage wheels.

(c) Chemical → heat → kinetic → electrical → kinetic.

Unit 4

125 By scale drawing find the magnitude of the resultant of the two forces shown.

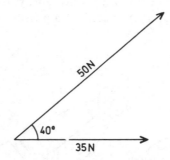

(East Anglian Examinations Board (N))

Complete the parallelogram and draw the diagonal from the 40° angle. Measure its length and convert to newtons using the scale of the diagram.

Unit 5

126 State the principle of moments. Describe how this principle may be verified for a system of parallel forces.

A uniform metre rule is pivoted at its centre on a horizontal needle so that it is free to rotate in a vertical plane. A weight of 10 N is suspended from the 0 cm mark, and another of 50 N from the 70 cm mark. Calculate the resultant moment of the forces on the ruler about the pivot when the ruler is momentarily at rest in the horizontal position.

What force must act at the 25 cm mark at an angle of 90° to the ruler in order to keep the ruler horizontal?

See unit 5.1.

Calculate the moments of the forces of 10 N and 50 N acting at 50 cm and 20 cm respectively from the pivot. The difference between these two moments gives the resultant, as they are acting in opposite senses.

Apply the principle of moments again, but this time with an unknown force acting at a distance of 25 cm from the pivot in the same sense as the 10 N force.

210

127 (a)

(i) A ladder has 15 rungs which are spaced out evenly along its length and the sides are made of two parallel pieces of wood of even thickness throughout their length. If you had to carry the ladder on your own, at which point would you hold it? Explain why.

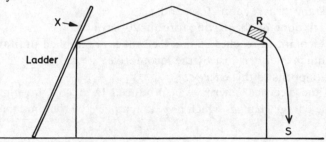

(ii) If the ladder was placed up against a building as indicated in the diagram above, why would it be dangerous to climb up to the point marked *X*?

(iii) A large block of stone is placed on the roof at *R*.
(*a*) What type of energy does it possess at *R*?
(*b*) What forms of energy exist immediately after it hits the ground at *S*?

(South East Regional Examinations Board)

(i) Hold the ladder at the centre of mass – halfway along. The clockwise and anti-clockwise moments about this point are equal and so the ladder remains horizontal.

(ii) It is possible that the moment of the person's weight about the edge of the roof would cause the ladder to rotate clockwise thus causing the foot of the ladder to lift from the ground.

(iii) Potential energy is finally transformed into heat and sound.

128 (a) The three diagrams below illustrate a toy which always returns to the upright position when released. On diagram *A*, indicate by a letter *X* a possible position of the centre of gravity of the toy.

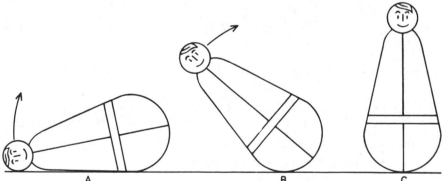

The centre of mass must be to the right of a vertical line through the point of contact of the toy with the ground in *A* and *B*. If this is so the toy's weight exerts a clockwise moment which rotates it into an upright position.

Unit 7

129

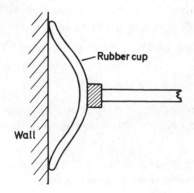

The figure shows part of a rubber sucker which is attached to the wall. Explain why it stays against the wall.   (*Middlesex Regional Examining Board*)

There is little air between the wall and the sucker as it is squeezed out on application of the sucker to the wall. Inside the sucker the pressure is low whereas atmospheric pressure acts outside. This pressure difference holds the sucker to the wall.

130 Draw a labelled diagram of an aneroid barometer and explain how it works.

See unit 7.5.

Unit 8

131 (*a*) State the law of flotation.

With the aid of a labelled diagram, describe a common hydrometer designed to measure relative densities between 0.9 and 1.1.

Describe how you would use the hydrometer to measure the relative density of sea-water.

(*b*) When a mass of 300 g is suspended from the 0 cm mark on a uniform metre rule, the rule balances horizontally on a knife-edge placed under the 20 cm mark. What is the mass of the rule?

The knife-edge is now moved to the 30 cm mark, but the 300 g mass remains at the 0 cm mark. Where must a mass of 250 g be placed to again balance the rule horizontally?

(*a*) See unit 8.4.
See unit 8.5. The inner weighted glass bulb and stem in Fig. 8.3 form the basic hydrometer. The mark 0.9 would be near the top of the stem and 1.1 towards the bottom.

(*b*) The rule balances 30 cm from its centre of mass. At balance the moment of the unknown mass of the rule at a distance of 30 cm from the pivot equals the moment of the weight of the 300 g mass at a distance of 20 cm. Moments are now taken about the new position of the knife-edge. The moment of the 250 g mass acts in the same sense as the moment due to the weight of the rule acting at the centre of the mass.

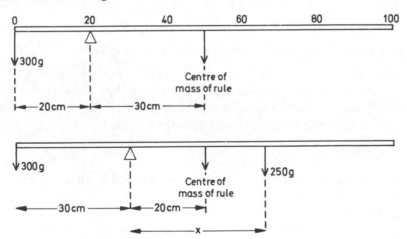

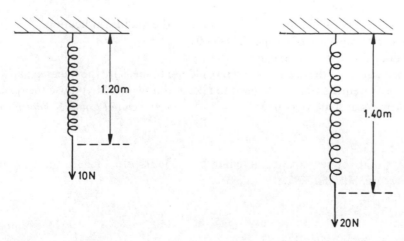

Unit 9

132 In the figures the length of the spring with 10 N hung on it is 1.20 m and with 20 N 1.40 m.

(a) What would be the length of the spring with 12 N hung on it, if the spring obeys Hooke's law?

(b) State Hooke's law. *(Middlesex Regional Examining Board)*

(a) If the spring obeys Hooke's law the increase in length is proportional to the force applied. 10 N causes an increase in length of 20 cm. 2 N causes an increase of 4 cm.

(b) See unit 9.1.

Unit 10

133 Complete the figure to show the water levels in the two glass tubes. *(East Anglian Examinations Board (N))*

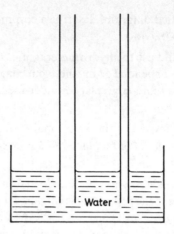

The weight of water supported is proportional to $r^2 l$, where l is the length of the water column.

The surface tension force is proportional to r. As the surface tension force always equals the weight supported, if r doubles, l must halve. The water is thus lower in the wider tube.

Unit 11

134 Use the kinetic theory to explain why a puddle dries up on a windy day.

Those molecules which have considerably more than the average kinetic energy of the molecules in the puddle, and are moving upwards near the surface, escape. On a windy day the molecules which evaporate are removed from the vicinity of the puddle by the wind and more escape and take their place. Evaporation thus continues until the puddle dries up.

Unit 12

135 A bimetal strip (compound bar) is straight when cold but bends when it is heated.
(a) Name two suitable metals for a bimetal strip.
(b) Of the metals you have chosen which expands more when heated?
(c) Give two uses of a bimetal strip. *(Associated Lancs Schools Examining Board)*

See unit 12.1.

136 (a) Describe an experiment that shows that a liquid expands on heating.
(b) The linear expansivity of copper is 0.000017/K.
What does this statement mean?
(c) Draw a diagram of the apparatus you would use to measure the linear expansivity of copper.
(d) What measurement would you need to take and how would you use the results to calculate the linear expansivity of copper? *(Associated Lancs School Examining Board)*

(a) See unit 12.3.
(b) It means that copper expands in length by 0.000017 of its length for each degree K rise in temperature. See unit 12.2.
(c) See unit 12.2.
(d) See unit 12.2.

137 Draw and label a clinical thermometer.

See unit 12.4.

Unit 15

138 Why is it that 1 g of steam at 100°C would cause a more serious scald than 1 g of water at 100°C? *(East Anglian Examinations Board (N))*

In changing to water at 100°C the steam gives out its latent heat (2.26×10^3 J). This is absorbed by the body as well as the specific heat due to the water cooling (265 J/K approximately).

139 Heat supplied steadily at the rate of 120 W to 0.6 kg of crushed ice raises its temperature from $-20°C$ to 0°C in 200 s.
   (*a*) What is the total heat capacity of ice?
   (*b*) What is the specific heat capacity of the ice?
   (*c*) Taking the specific latent heat of fusion of ice to be 336 000 J/kg, find how long, at the same rate of heat supply, it would take to melt all the ice at 0°C.
      Indicate briefly how heat might be supplied steadily to the ice and how the rate of heat supply could be measured.

   (*a*) The total heat capacity is the heat absorbed for each degree rise in temperature; that is it
   equals $\dfrac{\text{power} \times \text{time}}{\text{rise in temperature}}$
   (*b*) The specific heat capacity refers to 1 kg of ice and is calculated by dividing the value for (*a*) by the mass of ice.
   (*c*) Use the equation at the end of unit 15.3 to calculate *t*.
      See the diagram in unit 15.3.

Unit 16

140 Name **two** good conductors of heat and **two** heat insulators.
   See unit 16.1.

141 (*a*) (i) Two identical tins, one painted black and the other white, each have the same volume of water poured into them and are placed at equal distances from a source of heat. The temperature of the water in each tin is noted with the following results:

Time in minutes	0	10	20	30
Black tin temperature (°C)	20	23	26	29
White tin temperature (°C)	20	21	22	23

   Plot two graphs of these results on the same axis with the time on the horizontal axis and the temperature on the vertical axis.
   (ii) Explain why there is a difference in the temperature of the water in the tins at the end of the experiment.
   (iii) What will be the difference between the temperatures of the water in the two tins after 23 minutes? *(South East Regional Examinations Board)*

   (*a*) (i) ——————
   (ii) A black surface absorbs a higher proportion of the heat falling on it than a white surface. The water in the black painted tin thus receives heat at a faster rate than the water in the white one.
   (iii) Draw a vertical line through 23 minutes on the time axis. Draw horizontal lines through the points of intersection of the vertical line and the graphs. Read off the values where the horizontal lines cut the temperature axis. Find the difference in these values.

142 Explain the following in terms of conduction, convection and radiation of heat.
   (*a*) Houses in hot sunny countries traditionally have thick, white painted walls, with small windows.
   (*b*) A vacuum flask keeps tea hot.

(*c*) Glider pilots are able to find upward moving currents of air over certain features such as towns.

(*d*) The blackboard in a class room stays at the same temperature as the rest of the room, even though it absorbs more radiant heat-energy than the light coloured walls.

(*e*) On a cool day iron railings feel colder to your touch than a wooden bench.

(East Anglian Examinations Board (N))

(*a*) Thick walls reduce the heat which enters the house as a result of conduction through the walls. If the walls are white a small fraction of the energy falling on them is absorbed, most is reflected. Small windows mean that little heat enters by radiation.

(*b*) The vacuum between the walls reduces the energy entering by conduction or convection, as both these require a material through which to travel. Silvering of the walls facing the vacuum reduces the radiation which crosses this gap.

(*c*) Warm air has a lower density than cold air and hence rises. The gliders rise on the warm convection currents so caused.

(*d*) The blackboard absorbs more radiation from the surroundings than the white walls, but also radiates more. The temperature it acquires depends on a balance between its absorption and radiation and this is the same temperature at which the absorption and radiation of the white walls balance.

(*e*) Heat is conducted from the hands to the interior of the rails rapidly and this gives the sensation of cold. For wood the conduction is considerably less and hence the sensation of cold is less, making the wood feel warmer.

Unit 21

143 What is meant by *resonance*? State *one* practical example where this phenomenon is useful.

See unit 21.8.

Unit 22

144 Explain why a compass needle sets in an approximately North–South direction.

A compass needle is a small magnet. The magnetic field of the Earth compels one end of the needle to point in an approximately northerly direction. This end is called the *N* pole of the compass needle.

145

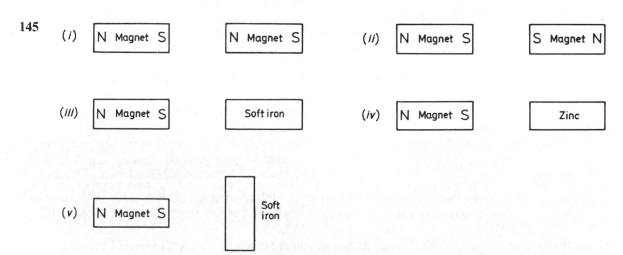

Indicate what would be observed between each pair of objects.

The magnetic fields for arrangements (*i*) and (*ii*) are shown in unit 22.3. In (*iii*) the soft iron becomes magnetised and the field pattern is similar to (*i*). Zinc is non-magnetic and so pattern (*iv*) is that due to the single magnet. In (*v*) the soft iron bar becomes magnetised with poles on its sides, *N* towards the magnet and *S* away from it. There is a strong field in the region between the magnet and soft iron.

Unit 24

146 (*a*) (i) Why would it be wrong to fit an electric fire in a bathroom on the wall directly above the bath?

(ii) Where should such a heater be fitted and what type of switch should be used to operate it?

(*b*) The flex to the 13 A 3-pin plug shown in the diagram has been incorrectly fitted. List four mistakes that have been made.

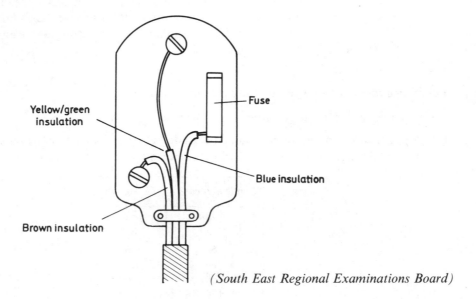

(South East Regional Examinations Board)

(*a*) (i) In this position it would be possible for someone standing in the bath water to grasp the live part of the fire, for example the element. They are also likely to have wet hands and thus to make good contact with the element and with earth through the water.

(ii) It should be placed on a wall out of reach from the bath and should be operated using a switch with a hanging cord.

(*b*) (i) The live lead (brown) should be connected to the fuse, whereas here the neutral lead (blue) has been connected to it.

(ii) In any case the blue lead has been connected to the wrong end of the fuse. As connected here the lead goes directly to the pin, and the fuse is not in the circuit.

(iii) Too much insulation has been removed from the earth lead. There is a danger of it 'shorting' to the brown lead.

(iv) The insulation round the entire cable has been stripped back too far. It should go under the cable clamp. As shown here any pull on the cable is taken by the three leads.

Unit 26

147 (*a*) Describe with the help of a labelled diagram the working of a simple dynamo. State whether the dynamo you have described produces alternating or direct current.

(*b*) (i) Explain how transformers are used in the National Grid system.

(ii) Why does this system use alternating current?

(East Anglian Examinations Board (N))

(*a*) See units 26.2 and 26.3.

(*b*) (i) See unit 26.6.

(ii) A voltage is induced in the secondary coil only if the magnetic flux through the coil changes. This changing flux is produced by a changing voltage in the primary coil. An alternating voltage is a convenient form of changing voltage as it can be obtained from a simple dynamo (unit 26.3). A steady voltage in the primary produces a steady magnetic flux and no voltage is induced in the secondary.

Unit 27

148 (*a*) Draw and label a diagram showing the structure of a transformer.
What are the two principal causes of energy loss? How are they minimised?

(*b*) Why cannot a transformer be used with d.c. supply?
What would happen if a high d.c. voltage were connected to the primary of a transformer?

(*c*) Draw a clear circuit diagram to show how a transformer and a diode valve may be used to produce low-voltage half-wave rectified output from the a.c. mains supply.

(*a*) See unit 26.5.

(*b*) A transformer relies on the fact that a changing magnetic flux through the secondary coil induces a voltage in it. A d.c. voltage connected to the primary provides a steady magnetic flux not a changing one. There is thus no induced voltage in the secondary coil.

(*c*) See the end of unit 27.2 (Fig. 27.2).

Unit 28

149 Name *three* methods of detecting ionising radiation.

See unit 28.1. The Geiger–Muller tube, the cloud chamber and the ionisation chamber.

150 A geiger tube samples some radioactive Radon gas, an alpha particle emitter, every 30 seconds. The results are given in the table below. Background count has been allowed for.

Time in seconds	Count rate
0	8040
30	5040
60	3360
90	2240
120	1440
150	880
180	600
210	400
240	240

(*a*) Draw a graph of count rate (vertical axis) against time.

(*b*) From your graph determine the time when
  (i) the count rate was 6000
  (ii) the count rate was 3000

(*c*) (i) Calculate the time which elapsed between the 6000 and 3000 count rates. (Use your answers to (*b*) above.)
  (ii) What is the significance of this time?

(*d*) What is meant by the term background count, and how is it caused?

(East Anglian Examinations Board (N))

(*a*) ———————

(*b*) (i) Draw a horizontal line through 6000 on the vertical axis. Draw a vertical line through the point where the horizontal line cuts the graph. The point where the vertical line cuts the horizontal axis gives the value required.
  (ii) Repeat the above procedure starting at 3000 on the vertical axis.

(*c*) The count rate is halved between (*b*) (i) and (*b*) (ii). The time which elapses between these two count rates is thus the half-life.

(*d*) The background count is the count which is obtained without a radioactive source in the vicinity. It is caused by cosmic rays in the atmosphere passing through the Geiger tube.

NUMERICAL ANSWERS TO LONGER QUESTIONS
GCE answers

 1 490 cm/s^2; $g = 9.80$ m/s^2

 2 (*b*) (i) 1.25 m/s^2 (ii) 625 N

 3 30 N;   21 m.

 4 (*e*) 0.5 s   (*f*) 0.75 m.

6 (i) 5 m/s   (ii) In the original direction of travel of the 200 kg body.

7 (b) One third.

  (c) (iv) The kinetic energy of P is half that of Q.

    (v) The speed of the combined trolleys is twice as great as in (b).

8 (a) (i) 225 000 J.   (ii) 30 000 kg m/s.

  (b) (i) -1.25 m/s^2.   (ii) 2500 N.

  (c) 112.5 m.

9 (a) 0.4 m/s.   (f) 1.6 N.

  (b) 0.16 m/s.   (g) 1.6 N.

  (d) 0.3 s.   (h) 3.0 kg.

  (e) 0.8 m/s^2 (i) 0.135 J.

10 (c) 12 m.

  (d) 42 m.

  (g) 20 m.

11 (a) 1/200th i.e. 3.6×10^4 N.

  (b) The force opposing motion is 3.6×10^4 N.

  (c) 4×10^5 J/s.

  (d) 7.2×10^4 N.

  (e) 8×10^5 W.

12 (b) 20.62 km/hr.

  (d) (i) $2\frac{1}{3}$ m/s in their common direction.

    (ii) $\frac{1}{3}$ m/s in the initial direction of the 1.0 kg trolley.

13 (a) (i) 48°36′ upstream.   (ii) 45 minutes 21s.   (iii) 1.5 km.

  (b) 3 m/s;   6 m/s.

14 (a) 100 N.   (b) 195 N.

15 (a) 18 N.   (b) 36 N.

16 (a) 173 N horizontally; 100 N vertically.

  (b) 173 N.

  (c) 2900 N.

  (d) 69.2 Nm.

  (e) 59.2 Nm.

19 (b) (i) 37 m/s.   (ii) 132 m.

20 (b) 15 N.

22 (i) 7.5.   (ii) 75 W.

  (a) 125 W.   (b) 100 m.

24 1.25.

25 (b) 1800 J.

26 (b) (i) 700 N.   (ii) 2100 N.   (iii) 10.5.

28 0.75.

30 108 kg; 600 N.

38 (d) 960 kg.

41 (a) 20 m^3.

42 0.81 cm^3.

43 (b) 477°C.

44 (c) 900 cm^3.

45 21.6°C per minute. 4167 J/kg°C.

47 (b) 480°C.

48 (b) (i) 2.4×10^5 J/kg°C; 3.88×10^5 J/kg°C. _____

    (ii) 15°C.   (iii) 80°C.   (iv) 2.16×10^7 J/kg.   (v) _____

49 0.6 g/s or 6×10^{-4} kg/s.

60 (a) 15 cm behind the mirror

  (b) (ii) A virtual, erect and magnified image 30 cm behind the mirror.

    (iii) A virtual, erect and diminished image 10 cm behind the mirror.

61 Refractive index = 1.52.

62 $C = 38°41'$.

63 In Fig. (i) the angle of refraction is 48°36'.
The ray emerges along the face of the block when the angle of incidence is 41°49'.

64 (b) (ii) 12.5 cm.

65 100 mm; 10 mm towards the object.

66 (b) An erect, virtual image, height 1.125 m, position 3.75 cm from the lens.

68 An inverted, real image, height 4.0 cm, position $9\frac{1}{3}$ cm from the lens.

69 20 mm in both cases.

70 (b) (i) 1 cm below the axis at I.

71 (b) 12.5 cm.

72 (a) 20 cm.

73 52.6 mm.

75 (a) (iv) 20 cm/s.

82 320 m/s.

84 (b) 3×10^8 m/s.

86 (a) 338 m/s.

91 10^{17} every second.

92 (a) (i) 18 ohms.   (ii) $\frac{1}{3}$ A.   (iii) 9 ohms.   (iv) $\frac{2}{3}$ A.
(v) 2 V.   (vi) 4 V.   (vii) 2 V.
(b) AD. $2\frac{2}{3}$ W.

93 0.5 ohms.

94 2.5 ohms; 0.4 A.

95 3.6 ohms and 2.4 ohms; 24 W.

99 Connect a 99 980 ohm resistor in series with the meter.

100 Connect a 195 ohm resistor in series with the meter.

101 (i) A 0.015 ohm resistor should be connected in parallel with the meter.
(ii) A $328\frac{1}{3}$ ohm resistor should be connected in series with the meter.

102 (b) (i) 4 A.   (ii) 60 ohms. A 5 A fuse would be suitable.

104 (b) (i) 12 V.   (ii) 2 A.   (iii) 16 W.   (iv) 36 W.   (v) $44\frac{4}{9}\%$.

120 4 minutes.

121 (b) (i) 20 g.   (ii) 10 g.   (iii) 2.5 g.

CSE answers

122 (b) (i) 0.5 m/s²   (ii) 10 950 m   (iii) $24\frac{1}{3}$ m/s

124 (a) (i) 1.25 m/s².   (ii) 62 500 N   (iii) 250 m   (iv) 15 625 000 J.
(b) (i) 15 625 000 J.

125 80 N at an angle of 10°52' to the force of 35 N.

126 500 Nm;   20 N.

131 (b) 200 g; at the centre of mass.

132 1.24 m.

139 (a) 1200 J/°C.   (b) 2000 J/kg°C.   (c) 28 minutes.

Index